CLASSICS
IN BIOLOGY

CLASSICS
IN BIOLOGY

A Course of Selected Reading
by Authorities

Particular attention is directed to the
Introductory Reading Guide by
Sɪʀ S. ZUCKERMAN, C.B., F.R.S.

KENNIKAT PRESS
Port Washington, N. Y./London

CLASSICS IN BIOLOGY

Copyright, 1960, Cultural Publications (I.U.S.), Ltd.
Reissued in 1971 by Kennikat Press by arrangement
with Philosophical Library, Inc.
Library of Congress Catalog Card No: 78-122974
ISBN 0-8046-1355-9

Manufactured by Taylor Publishing Company   Dallas, Texas

ESSAY AND GENERAL LITERATURE INDEX REPRINT SERIES

CONTENTS

BOOK I
THE UNITY OF LIFE

v

CONTENTS

BOOK II

THE DIVERSITY OF LIFE

CONTENTS

INTRODUCTORY READING GUIDE

BY

Sir S. ZUCKERMAN, C.B., F.R.S.

LIVING AND NON-LIVING

THE environment in which we live is made up of living and non-living things. In the living world we include ourselves, and all animals and plants ; in the non-living, air, water, and (leaving out such things as bacteria and the roots of trees and plants) all the constituents of the soil. Even though there is usually no difficulty in distinguishing between what is alive and what is not, the borderline between the two is very indistinctly and vaguely drawn.

The Inanimate World. The inorganic and organic substances which compose the inanimate world are made up of molecules, each of which consists of a constellation of atoms. Examples of the inorganic are such things as air, iron, water, salt, and bricks ; of the organic, alcohol, acids like the acetic acid that occurs in vinegar, carbohydrates (e.g. sugar), fats, and protein. Fundamentally the difference between the inorganic and the organic is one of quantity, not quality—the organic molecule is the more complicated. Proteins are the only things whose molecular structure still remains a major mystery. The protein molecule is constructed out of simpler molecules called amino-acids. There are some twenty of these, and the unique characteristics of different proteins are due to their formation from different combinations and arrangements of these amino-acids. Proteins are the main constituents of flesh, but the form in which they exist in the living body is different from that of proteins when they have been extracted from meat or blood and kept as a dry powder in a bottle. In the body, they have the capacity of self-reduplication, or reproduction.

THE LIVING WORLD

The living world consists of organisms—animal and vegetable. Structurally, organisms consist of single cells or of constellations of cells, each of which in turn consists of a highly complex constellation of molecules making up what we call the cellular protoplasm. We may regard the human being as the most complex of all organisms (though admitting that it is only his brain that makes man more complex than, say, a dog or a horse). Our bodies consist of millions and millions of cells organized to form different tissues and organs. The human brain is an assemblage of cells. So is the liver, so is muscle, and so, in spite of its fluid matrix, is blood. These various assemblages of cells are specialized for different functions, and the unified activity of the body depends on their combination and co-operation. The heart is, as it were, meaningless in itself, but upon its activity depends the activity and life of all the other structures in the body—brain, eyes, muscles, and so on.

At the other end of the scale of living things are organisms which consist of a single cell. The most familiar example is a microscopic aquatic creature called " amoeba ". The shape of amoeba, which is approximately 1/500th of an inch in diameter, is both irregular and changeable. Its cell consists of an inner mass of granular fluid protoplasm bounded by a relatively clear and gelatinous sheath. Within the cell is a denser spheroidal body—the nucleus. This contains a tangle of thread-like specialized protoplasmic structures called chromosomes. All nuclei contain such chromosomes, though they are particularly difficult to see in amoeba itself.

The one-celled amoeba moves in its watery medium by throwing out blunt projections of protoplasm. The clear gelatinous sheath at the hinder end contracts and thrusts forward the inner granular protoplasm, which seems to break through the front end to form the blunt finger-like lobes called " pseudopodia ". The inner fluid " sets " to form a new gelatinous layer at the front end, and so the creature moves. Particles of food in the water are surrounded by the pseudopodia, and so become included in the cell, where they are broken down into their molecular constituents and used for repair, growth, reproduction, and energy. Waste products are pushed out of the cell or simply diffuse through its outer layer.

When amoeba reproduces itself, it simply divides into two. This so-called " binary fission " is preceded by the division of the nucleus, and that, in turn, by the very exact lengthwise division of each of its chromosomal threads. Each daughter cell thus obtains a full and equal set of chromosomes. No form of sexual reproduction has been recorded in this creature.

Simple though it is, amoeba has a much more complex organization than a bacterium. This is only a twentieth of the length of an amoeba and often less. Its shape does not change and its internal structure is very much more homogeneous. At a level of size and organization far below the bacterium is the virus.

Viruses. It is the virus which inhabits the hazy boundary zone between the animate and inanimate worlds. Viruses can be seen only under the very high magnifications (× 5000–× 20,000) made possible by the electron microscope. Some viruses apparently consist of a single giant protein molecule. Viruses possess some of the properties of non-living matter—for instance, when isolated, they can be crystallized. Yet within the cells of the higher animals on which they prey, viruses multiply their number at the expense of the host cell. Bacteria can live and multiply indefinitely outside cells—in, for example, the special but not particularly complex culture-media in which the bacteriologist keeps them. But no one has yet grown a virus or even persuaded it to show fermentative activity outside a living cell.

Whether or not the viruses can be said to belong to the living world is purely a matter of definition. They have no independent activity, and they can only be recognized as viruses—that is to say, they only behave as such—when they are inside the cells of some organism. If they are forms of life, they are parasitic and completely degenerate forms : bacteria, perhaps, so degenerate that only the faculty of self-reproduction has been retained.

If it were ever possible to synthesize a protein molecule out of a mixture of synthetic amino-acids, and to cause it to renew itself and reconstruct its own pattern outside the environment of the body of some living thing, then the barrier between the living and non-living worlds, as we now understand it, would be destroyed once and for all.

The Instability of Cellular Structure. Leaving viruses on one side, all living cells contain fats or fatty acids, alcohols, aldehydes, carbohydrates like sugar, starch or cellulose, and so on. All contain inorganic salts as well—significantly enough, the salts that occur in sea-water. Compounds of phosphoric acid are of particular importance, and the combination of organic compounds with phosphoric acid is an important intermediate step in many chemical transformations of the cell. Proteins also occur universally in living things ; but whereas glucose alcohols, and phosphoric acid are chemically the same in whatever organism one may find them, the proteins are all different. One human being has a different protein make-up from another human being, and a very different make-up from a mouse. The pervasiveness of proteins and the multitude of different chemical reactions and transformations which their different forms make possible gives some point to the claim that life is a mode of the activity of protein molecules.

Yet although tens of thousands of chemically different proteins exist, all are built up of different combinations in different proportions of the same small number of amino-acids.

How do such complex systems as living things endure in time? Modern physics has shown that even the simplest inorganic substances are unstable. The molecule is made up of a constellation of atoms and each atom consists of a centrally placed and positively charged nucleus round which one or more negatively charged electrons revolve. The nucleus accounts for little of the volume, but most of the mass of the atom, and consists of positively charged protons and of neutrons which, as their name implies, are electrically neutral. Spontaneous changes take place in some atoms as a result of the loss of these particles, and in this way a " radio-active " substance such as radium eventually becomes degraded into lead.

If the internal structure of the inorganic world is unstable in the sense this description implies, that of the living world is even more so. The body consists, as we have seen, of cells organized to form different tissues and organs—e.g. blood-vessels, nerves, muscle, the liver, bone. Apart from the changes in size and shape which occur during growth, the form of the body is constant. But at any given instant its components are changing. Part of this change is due to wear. The cells of the soles of the feet and the lining of the gut wear away and are replaced. Another part is due to natural physiological turn-over. The cells of the liver become charged with the products of digestion, brought to them in the blood by which they are bathed, and at a later stage they give back to the blood nutrients for transport to other parts of the body where they are oxidized or burnt, so providing energy and heat. There is a steady small loss of nitrogenous compounds that must be made good. Water and salts must also be replaced.

But the body is in even greater flux than this; it also undergoes a turn-over at the atomic level. This fact has emerged from recent studies in which living things have been fed on food containing radio-active or on naturally occurring but rare isotopes (heavy carbon, heavy oxygen, radio-phosphorus, etc.) whose movement from one to another part of the body can be followed by means of Geiger counters. The body does not discriminate between an element and its isotope, and uses either indifferently. These researches with " labelled " atoms have shown that the body is in a dynamic, not a static, condition; only its *form*, or more generally its *organization*, is static. The constituent physical elements of its tissue, even the calcium and phosphorus-containing materials of teeth or bones, are not fixed once and for all in the places where they are to be found at any given moment. They are constantly on the move, and their places in the constellation of which they are part are immediately taken by incoming particles from the environment. It is like a tank into which water pours as fast as it flows out. The level is always the same ; the tank is always as full ; but the molecules of water in the tank are continually changing.

Characteristics of Living Things. In so far as flux is a characteristic of both living and non-living matter, their self-maintenance can only be thought of in relation to the immediate environment in which we find them. But here we can see what appears to be a major difference. The inanimate thing is not dependent on its environment for the maintenance of its inanimate state. We could, as it were, imagine a lump of lead continuing as such in a vacuum. In isolation, on the other hand, the living thing ceases to exist as such and disintegrates. The persistence of individuality, or more correctly, of individualized activity under certain specified conditions, is thus a *sine qua non* of living. So, too, are growth, movement, and reproduction. Many living tissues, skin for one, can be kept in " suspended animation " at the temperature of liquid air. Only then are they comparable with inorganic objects.

However complex the organism, the chemical elements found in it are the same as those found outside it. We know, too, that the physical law of Conservation of Energy applies as absolutely and rigorously to the living creature as to any lifeless machine, and that the working of the living body can be usefully studied on the basis of mechanical principles. The body needs a supply of energy, which it can neither create nor destroy, in order to make it go. Muscle is a form of internal combustion engine, rather more efficient than a steam locomotive and rather less efficient than a car ; so much muscular work needs the combustion of so much fuel. The major difference between living and non-living things, so far as the transformation of energy is concerned, is this. In the inanimate world, as in steam engines and cars, energy exchanges are accompanied by considerable alterations in temperature. Thus in order to release the energy from a lump of coal, it is necessary to combine the coal with oxygen. In so doing a fire is caused and we end up with energy in the form of heat and expanding gas. In the living thing, any energy transference which was associated with a big change of temperature would be incompatible with continued life. Nearly all animals and plants are killed by temperatures higher than 110° F. The basic chemical problem of all vital transformations, it has been said, is the achievement of chemical changes in an isothermal medium, by which large amounts of energy can be made use of in the small steps which alone are permissible in such a system.

All these attributes of living things are manifestations of catalytic chemical processes which occur within a constant narrow range of temperatures on the surface of certain cell proteins, called enzymes. Enzymes are indispensable to the transformations by which food is ingested and on the one hand broken down to provide the energy that makes the body's processes possible, and on the other built up to provide for the growth and repair of the body's tissues. The self-maintenance of bodily shape and function thus implies a continual sequence of chemical processes of which the assimilation of oxygen and of food is an essential part. This so-called " metabolism " means the assimilation of energy in some form or other, and the transformation of this energy into the activities of growth, repair, reproduction, movement, heat-formation, and so on.

It is not only in this way that the living thing is dependent on its environment. It also has sense organs, by which it perceives changes in the world around it—the sequence of day and night, the appreciation of heat, salinity, acidity, movement, and so on. In a certain sense an inanimate object may also be sensitive to changes in its environment. A thermometer, for example, reacts in a specific way to a given change in its environment. The " perception " of the thermometer, however, is not followed by any particular change in its behaviour; whereas perception in a living organism is, as it were, a signal for it to change its orientation within or adjustment towards its environment. Perception, as applied to living organisms, is a special case of " sensitivity ", which is common to both the living and non-living world.

THE EVOLUTION OF LIVING THINGS

The earth's surface and soil, the sea, sea-bed, and sky contain millions of different sorts of creatures which we call alive. Of insects alone, about a million different sorts are known. We recognize two vast kingdoms of living things, vegetable and animal. The plant is no less " alive " than the animal. Its mode of life may be less conspicuously active, and may thus deceive us if we regard spontaneous motion as the mark of life; but, in fact, the plant has unique powers, such that the whole animal kingdom, including man, depends upon them. By virtue of the green matter called chlorophyll in its leaves, it is able to make use of the energy of sunlight for the synthesis of nutrients. The dead bodies of animals which have lived, directly or indirectly, upon plant nutrients, are in their turn eventually used by plants. There is thus a cycle of life. When we burn coal, which is fossil wood, we are also completing a cycle, in so far as we then release the energy of the sun that was stored in the tree during its life.

What is known about the virus suggests that it has a homogeneous structure. The protoplasm of the single-celled amoeba, as we have seen, is differentiated into nucleus and outer and inner layers. At the next stage of life cells join together to form a colony, but every cell within the colony can perform such functions as digestion and reproduction, and is to this extent independent of its fellows. At the next stage different cells within what is in effect a single organism are set aside for different functions—defence, reproduction, and co-ordination —and acquire correspondingly different structures. In much the same way, although we are derived embryologically from a single fertilized cell, our body differentiates during its development into a system of blood-vessels, a nervous system, a muscle system, an alimentary system, a reproductive system, and so on.

We know that every living thing is derived from a living parent or parents : " *Omne Vivum ex Vivo* ". The alternative belief that living creatures may develop without a living progenitor—that is to say, the belief in " spontaneous generation "—was at one time widely held. But during the last century it was the subject of prolonged study and controversy, and Louis Pasteur in France, and Tyndall and T. H. Huxley in England, all demonstrated that every supposed piece of evidence of spontaneous generation was false. No evidence of the creation of life from non-living matter has emerged since their time. We do not, of course, look for evidence of the spontaneous generation of elephants or oaks ; but it is essential to realize that evidence cannot even be found of the spontaneous generation of the most minute microbes or viruses.

By common consent the inanimate world has had a longer cosmic history than the animate, and while many are satisfied to assume that the living protein molecule evolved from non-living molecules, others find it necessary to invoke the operation of a divine power to explain the emergence of life. Whatever view one takes, it is essential to remember two things. The first is that it would be wrong to suppose that our planet, before life began, was just as it is to-day, only stripped of all living things. The earth has been transformed by the very existence of life. For example, it is hardly probable that oxygen existed in free molecular form before life began. Moreover, the fact that the environment of the world is to-day suited to the existence of living things can be regarded as a consequence of the interaction of life and the earth in their mutual evolution.

The second thing one should bear in mind is that the mystery of the origin of life and its evolution is no greater than that which has to be unravelled in attempting to discover how living things work. The evolution of man from organisms as simple as *Amoeba* is no less or more mysterious than his development from a single fertilized egg-cell of much the same size and degree of visible complexity. Familiarity deceives us into taking the latter process for granted.

Geological Time. Whatever the origin of life, its later history may be read, quite plainly, in the record of the rocks. Here the study of geology, primarily concerned with the mechanical and chemical forces that have raised the mountains and hollowed out the ocean basins, acquires a new interest. The fossil remains of living things, dating back some 600 million years, tell us that in past ages other forms of life than those now extant lived upon our globe, and that there has been a process of change, of transformation and diversification, in which some forms of life have disappeared, and other new ones appeared. This process is called evolution.

The Earth consists of a molten core surrounded by a crust which has slowly solidified through the ages. The more recent part of its history, that in which life evolved, can be subdivided into geological phases. We live in what is called the recent or Holocene epoch, which began 25,000 years ago. This was preceded by a million years of what we call the Pleistocene epoch, the Pleistocene being the start of the Quaternary Period of the world's history. Before that was the Tertiary period, lasting about 75 million years, and consisting, as we go backwards, of the Pliocene, the Miocene, the Oligocene, and the Eocene epochs. The Quaternary and Tertiary constitute the Cainozoic or modern-life era. It was preceded by a Mesozoic, or middle-life, era, lasting about 130 million years, and comprising periods called Cretaceous, Jurassic, and Triassic. This in turn was preceded by the Palaeozoic, or ancient-life, era, lasting about 300 million years and subdivided into the Permian, the Carboniferous, the Devonian, the Silurian, the Ordovician, and the Cambrian. The Palaeozoic era does not, however, represent the birth of our globe. It takes us only some 500 million years back, and was preceded by the Archaeozoic or primeval-life, and the Azoic or lifeless eras. The minimum age of the earth, as determined from radio-activity estimates of terrestrial material, is 2000 million years.

The first evidence of animal life is in early Cambrian deposits, which were laid down, according to estimates made by a variety of geological techniques, including radio-activity studies, about 500 million years ago. Living things, we can hardly doubt, had existed millions of years before, but the earliest organisms, things like protozoa and algae, did not leave their remains as fossils. Among the animal forms we find in Cambrian formations are shell-fish and worms, and the fossil remains of scorpions go back to the Silurian, 350 million years ago.

The Progression of Animal Life. The major subdivisions of the animal kingdom—if we exclude the sub-kingdom of Protozoa in which are classified all the single-celled organisms like *Amoeba*—are the Coelenterates, which include things like jelly-fish, sea-anemones, *Hydra*, and the Portuguese Man-o'-War ; the Platyhelminths or flat-worms, including also such parasites as the tape-worm and liver-fluke ; the Annelids or true worms, like the earth-worm itself; the Arthropods, containing creatures like centipedes, scorpions, crabs, insects, and spiders ; the Nematodes or round worms, many of them parasites ; molluscs or true shell-fish such as oysters, snails, and octopuses ; Echinoderms, among which are classified sea-stars and sea-urchins ; and the Chordates, the creatures with a backbone, however primitive, and (except in the lowest forms) an internal skeleton. Representatives of almost all these major groups of the animal kingdom are present in Ordovician formations, that is to say, in geological deposits reaching back nearly 400 million years. The first of the Chordates, the group of organisms to which we belong, were aquatic, jawless fishes, probably similar to the modern lampreys. During the succeeding Silurian and Devonian periods other types of fishes arose and many of their modified descendants survive to this day.

The next group of chordates to appear were amphibians, such as frogs, newts, and toads. They were almost the first vertebrate animals that were able to leave the water and live on dry land for a part of their life-cycle—not quite the first, because the lung-fish, a natural link between fish and frog, can do so too. They appeared during the Devonian, 300 million years ago, but just as they do to-day, even the most terrestrial and dry-skinned of them, e.g. toads, had to return to the water to breed. Reptiles were the next chordates to emerge, and the first completely non-aquatic vertebrates, for they laid eggs with shells on land. After their origin in the Carboniferous, some 250 million years ago, they rapidly blossomed into a wide variety of types which flourished throughout the whole of the Mesozoic era. Reptiles are now a dying class, but one group of these early reptiles gave rise to the birds, which first appeared in the Jurassic 120 million years ago, and another, about the same time or even earlier, to the mammals, which eventually became the dominant type during the Tertiary epoch. There still exist in Australia two very primitive mammalian forms, the Duck-billed Platypus and the Spiny Ant-eater, whose young are hatched

from the egg, like those of reptiles and birds. Both of them nevertheless
suckle their young in the same way as all mammals do.

Man belongs to a group of mammals, the Primates, which includes
apes and monkeys. Fossils that can be identified as members of this
group appear in early Eocene formations, nearly 70 million years ago.
Creatures akin to the monkeys and apes have existed since the beginning of the Oligocene, 40 million years ago. But there is no evidence
that man existed before the Pleistocene—less than a million years ago.
During these million years various types of primitive cavemen lived
in the Far East, in Africa, and in Europe, some of them differing
in form from modern man. Some were very similar in skeletal
structure to ourselves, but we do not yet know which were our ancestors,
or, indeed, where the line of descent leading back separated from that
leading to the monkeys and apes.

Before we leave the subject of the geological record, it is useful to
note that new animal forms did not appear in an orderly and even
sequence during the history of the earth. Evolution was often
" explosive ", in the sense that when a major new group of organisms
appeared, it very rapidly radiated into its major sub-groups, after which
a period of relative quiet intervened before yet another phase of rapid
change began. We must remember, however, that we are talking about
geological time—not time as we experience it year by year—and that
the term " rapid radiation " may encompass some millions of years,
during which the climate of the earth was subjected to violent change.
Life began, as we have seen, 500 million years ago or more, and man
started his history only about one million years ago. But in the
relatively short period our own human family has existed, the polar
ice-cap has several times advanced and receded in waves which have
covered the whole of Europe and Northern Africa. Forests have come
and gone ; continents have changed their shape ; and hot deserts have
taken the place of swamps. The back-screen against which new animal
forms have emerged in the world has been constantly changing.

Variation the Basis of Evolutionary Change. The evidence of the
geological record, coupled with the facts derived from studies of comparative anatomy and of the processes of development, have shown, as
clearly as one can expect, that the living things we know to-day have
evolved from simple beginnings. What we mean by the evidence of
the geological record is the appearance in successive geological epochs
of new and successively modified forms. By the evidence of comparative anatomy we mean that the structure of one organism is very
similar to that of another ; for example, the disposition of our organs,
muscles, and bones corresponds so closely with that in the apes and

monkeys that we classify ourselves with these animals in the Order of animals called Primates, to show that it is distinct, say, from another Order in which we classify all the carnivorous mammals. Such similarity of basic structure argues for a common ancestry. By the evidence of development we mean that animals, as they develop, pass through stages very closely similar to the developmental stages of lower forms. At one stage of its life a human being has gill slits like a fish ; later it develops a tail like a reptile ; later still, a pelt of hair like an ape. All these are lost as the definitive human pattern emerges. Unless these so-called " recapitulations " are gratuitously misleading, they too speak for an evolution of man from fish-like and latterly reptile-like and ape-like forms.

When *Amoeba* reproduces it just divides into two similar parts. In sexual reproduction, however, the offspring are slightly different from each other and from their parents. The slight inborn differences between such offspring are the material that evolution shapes. In sexual reproduction a germ cell from the male parent fuses with one from the female parent to form a single cell called the " zygote ". In other language, a male sperm fertilizes a female egg-cell or ovum. The fertilized cell or zygote divides and subdivides repeatedly, its daughter cells becoming differentiated to form the foundation cells of different bodily systems. At a very early stage of development some cells are set aside to form the germ cells for the succeeding generation.

It is in the union of the male and female germ cell that the possibility of recombining or " shuffling " the hereditary elements of the species exists. Each cell of the adult human body has a nucleus in which, as in the nucleus of *Amoeba*, are found pairs of protoplasmic threads, the " chromosomes ". The human body has twenty-four such pairs. Down the length of each chromosome are distributed particles called " genes ", the physical determinants of heredity. Genes are believed to control not only the working of the cells and tissues of the adult but also the way in which the fertilized egg develops.

The male sperm which fuses with the female ovum in sexual repro-duction transmits to the offspring genes or hereditary factors from the male parent ; the ovum transmits genes from the female parent. Since two germ cells unite in sexual reproduction, it is obvious that some process must operate to ensure that the zygote they form does not contain a double load of genes and of the chromosomes that carry them. The mechanism is revealed in the way the germ cells are formed. When the ordinary cells of the body divide, say the cells of the skin, they do so in such a way that each " daughter " cell receives an identical and full complement of genes. On the other hand when the germ cells are formed from the division of a precursor germ cell, the number of chromosomes, and hence the number of genes, becomes halved, each germ cell receiving a randomly allotted half-share of those in the parent cell. As no two individuals of any species are identical, i.e. no two have identical complements of genes, this mechanism ensures that genes from both parents, both grandparents, and so on, are randomly distributed among the offspring. Other more complex processes occur during the formation of the germ cells which help to ensure an even greater chance of recombinations of genes among the offspring. Further variability is introduced by occasional molecular changes or " mutations " in individual genes themselves. Although the frequency of mutation may be accelerated by some physical agents (e.g. X-rays and atomic radiation), its cause in the living animal is unknown.

Sexual reproduction thus not only enables the characters of one generation to be passed on to the next, but also ensures their reassortment and recombination in a virtually endless variety of ways. By unravelling these processes, modern genetics has furnished an explanation for the variability which is observed in all species of animals and plants, and has shown how this variation may be maintained or extended.

Selection. Zoologists have always been struck by the precise way in which animals are " fitted " to their environments. A horse is adapted for rapid movement over open grassland, and a monkey for movement in the trees. The degree of adaptation is sometimes so great that creatures become specialized to the point that their tolerance of environmental change is extremely limited. If their environments alter in any material respect, they die. Many parasites are specialized in this extreme way. Several explanations have been advanced to explain this phenomenon. The one which is now widely accepted was first proposed by Darwin, almost a century ago. Following Malthus's observations about the pressure of population on resources of food and living space, Darwin observed that more organisms are born than can survive, and therefore postulated that there is a perpetual struggle for existence. Since all animals vary, he concluded that some will be better fitted to survive than others, and having survived, to transmit their qualities to a next generation. This process is now called environmental selection, or more simply, selection. Modern genetics, as already noted, has provided an explanation of the variability that exists within each species of animal. Evolution is the process of environmental selection working on a varying population of organisms.

This explanation has received very strong support from laboratory and field experiments. Sometimes change may be rapid. For instance,

the teeth and skulls of a group of green monkeys which were transplanted 300 years ago from their home in West Africa to the West
Indian island of St. Kitts are now measurably larger than those of their
modern mainland descendants of the parent stock. The striking
changes that have occurred in domesticated animals are legitimate
examples of experimental evolution. If the selective conditions change,
and a group has insufficient available variability or " plasticity " to
respond, it becomes extinct. In this way, the mammoths of the ice-ages
and the vast reptiles of the Mesozoic epoch presumably disappeared
from the earth. If the animals we know continue to change, we shall
deduce that they are not only well adapted to their present environments,
but that they also maintain sufficient potential variability to respond to
further selective pressure.

The process of selection is thus determined by changes in the
physical and living environment. Trees disappear from a region of
forest, and the creatures which lived in them either move on to other
forests, or adapt themselves to a more arid existence on the ground, or
become extinct. Predatory animals suddenly become more numerous
in an area. Creatures on which they prey either respond by developing
new defences, or they move to another area where they are possibly
less suited, or they are all destroyed. In the world of to-day man
represents the major selective force. The vast changes that have
occurred in the fauna and flora of the world over the past few hundred
years—the disappearance of animal species and the destruction of
natural forest—have occurred as a consequence of the spread of human
communities and of the growth of human civilization. Man has also
deliberately altered the nature of the animals and plants he has domesticated. For example, by repeated breeding of wheat plants chosen
because of their high yield, it has been possible to evolve strains whose
average yield is higher than that of the type from which the original
plants were chosen. This process has also been effective in improving
such factors as egg-production in poultry, or milk-yield in cattle.
Unfortunately, if any one character is selected too intensively, undesirable changes may occasionally be produced accidentally in others.
In almost every instance where selection of a single character has been
carried out in the laboratory, the result has been associated with a
decline in fertility.

Eugenics. The fact that man has been able to do these things has
frequently encouraged people to think that, given the will, he would
also be able to control his own breeding, and so to evolve into a kind of
superman. This belief is the basis of the subject called eugenics.
Through the practice of eugenics, it is assumed that one could breed

mental defectives out of the community and eradicate weak physical types of various kinds. Unfortunately the problem is less simple than appears on the surface. In only a few cases do we know the hereditary background of human characteristics, and this ignorance imposes an initial limitation on attempts to apply the laws of heredity in order to " improve " the human stock. In the few clear-cut cases in which their genetic basis is known, hereditary diseases could in theory be eradicated—provided that one could control the breeding not only of the affected individuals but also of the unaffected " carriers " of defects whose inheritance is governed by recessive genes.[1] However, large-scale attempts to " improve " the human stock in the final analysis depend on individual views about the direction in which a long-term improvement or advance should occur. But as Julian Huxley points out : " the formulation of an agreed purpose for man as a whole will not be easy. There have been many attempts already. To-day we are experiencing the struggle between two opposed ideals—that of the subordination of the individual to the community, and that of his intrinsic superiority. Another struggle still in progress is between the idea of a purpose directed to a future life in a supernatural world, and one directed to progress in this existing world. Until such major conflicts are resolved, humanity can have no single major purpose, and progress can be but fitful and slow."

THE WORKING OF THE BODY

If we are unlikely to realize within many lifetimes the possibilities that lie in the conception of eugenics, man has certainly been able to do much to reduce the risks he runs from diseases of various kinds. Over the years we have been slowly gaining knowledge about the character of the human body and about the manner of its working. In 1628 Harvey discovered the fact of the circulation of the blood, from which stems our knowledge of the action of the heart and blood-vessels. In 1794 Galvani discovered that nerves convey electric currents which stimulate muscles to contract, and so laid the foundation of our know-

[1] " Recessive " genes are those which have no outward expression unless an animal has received one from *both* its parents. Unfortunately it can be proved that with rare genes the unaffected carriers (who have inherited the gene from only one parent and so give no evidence of its presence) enormously outnumber the affected victims—so that it is really quite impracticable either to identify the unaffected carriers or to hope to control their mating. Nor should it be forgotten that recessive genes like that for haemophilia may be expected to turn up " spontaneously " by mutation about once in every 50,000 human beings.

ledge of the nervous system. In 1849 Berthold, a young Alsatian doctor, discovered that some tissues of the body are specialized to produce chemical secretions which, when passed into the blood-stream, stimulate other organs to undergo changes. His particular observation was that the testes of the male bird secrete into the blood a chemical substance upon which the growth of the comb, wattles, and the male plumage depends. By so doing he provided the proof of what had only been suspected before—that in addition to the effects of nervous stimulation, the functions of the body are integrated by a system of chemical substances called hormones. Later came the modern science of nutrition, with its studies of nutrients, vitamins, and minerals, and also the wider studies of biological chemistry.

Germs and Disease. All these researches were partly animated by the belief that only through knowledge of the way the body functions could we hope to eliminate the various ills by which it is beset. They were paralleled by other enquiries into the nature of disease itself. Here we probably owe more to Louis Pasteur, a Frenchman who was born in 1822, and died in 1895, than to almost anyone else. He is known to-day as the father of preventive medicine. Given the microscope and his own original genius and amazing devotion, he evolved and established the " germ theory of disease ", according to which many of the ills that flesh is heir to are due to invasion or infection of the body by minute parasites, most of which belong to the biological category of bacteria. Bacteriology thus suggested the possibility of preventing disease, and preventive medicine was born. " It is in the power of man ", wrote Pasteur, " to make all parasitic diseases to disappear from the earth." Pasteur held no medical qualifications. As a young chemist he studied fermentation, and found that the fermentations which occur in the formation of beer are due to minute living creatures, and that unwanted contaminating microscopic creatures cause the undesired fermentations—the " diseases " of beer. Thence he proceeded to the study of diseases in the vine ; in the silkworm ; in sheep and cattle and dogs ; and finally in man. In every instance infection by bacteria and fungi, causing morbid fermentations, was the outstanding fact. It was the chemist and the biologist who became the greatest hygienist of all time. Round the base of the dignified and eloquent memorial to Pasteur erected to him in Paris by international subscription, we see the Vine saved from parasites and death, Sheep and Cattle thus saved, and under the sculptured benignity of the master's gaze, the form of a young girl saved from Death, a shrouded figure slinking away with his scythe to find some victim not protected by this veritable Shepherd of mankind.

To Pasteur and his followers, and to his great English forerunner, Edward Jenner, we owe numerous instances of the principle that diseases may be prevented by the careful education or forewarning of the body against them, by the deliberate introduction of the infective agent in an attenuated and relatively innocuous form. This method is still usually called Vaccination (Latin *vacca*, a cow), because Jenner used the modified form of human smallpox, called cowpox, that occurs in the cow as a source for his " vaccine matter ".

When the body is invaded by bacteria, it responds by building up a state of immunity. Whether these are harmful or not, all foreign proteins, for example those introduced by bacteria, provoke the body to manufacture a special complementary protein or " antibody " which destroys, neutralizes, or inactivates either the foreign immunity-provoking agent or the bacterial cell that carries it. Unfortunately there is a time-lag between the invasion of the foreign organism or bacterium and the production by the body of antibodies to match it. Meanwhile the bacteria may get out of hand, produce harmful toxins, or even multiply in the blood-stream, so that disease or death may result. The natural method of defence may, however, be assisted in two ways. First, an additional supply of antibody may be given from an animal which has already responded to the infection. This confers only a transient or " passive " immunity. This is what is done when the serum of someone who has had infantile paralysis, which is due to a virus, is injected into a person in the early stages of the disease. A more permanent, or " active " immunity can be produced by injecting doses of the disease-causing agents small enough not to prove fatal, but which are, however, adequate to stimulate the cells of the body to produce a continuous source of antibody, and to accelerate its production when subsequent invasion by bacteria takes place. The familiar process of vaccination makes use of this principle, as does the anti-typhoid injection.

The New Drugs. Immunizing the body against various bacterial invaders is only one of the new ways in which infective agents are countered. Once the invaders have taken hold modern science has other powerful means of dealing with them. The sulphonamides, or " sulpha drugs ", were discovered (in the form of Prontosil) by Professor G. Domagh in the 1930's, and have proved to be specific against several types of bacterial infection. The very nearly simultaneous discovery of penicillin by Sir Alexander Fleming, Sir Howard Florey, and Dr. Chain provided us with even more powerful weapons with which to counter disease. Penicillin is a chemical substance that is produced by a mould. Its discovery was due to the observation that cultures of bacteria failed to grow when they became accidentally infected with this particular mould, and it has opened the way to the discovery of other so-called antibiotics—streptomycin, aureomycin, chloromycetin, and others, each of which may have a specific part to play in the battle against infective disease.

Nutrition. But it is not only in learning how to counter bacterial invaders that we have leapt ahead in our control of disease. Many illnesses are " deficiency diseases " resulting from malnutrition. The good health enjoyed by the crowded cities of the modern world is, in fact, due as much to our understanding of what the nutritional requirements of the body are as it is to the sanitary laws that have to be enforced if we are to avoid the ravages which could be caused by the spread of bacteria. We measure nutritional levels in terms of calories, and we know that when their number falls below, say, 2500 a day, under-nutrition is being approached. People maintained at such a nutritional level have less energy to expend than people on higher diets. We know, too, that calories are not the only things that matter : they are merely the measure of the energy provided by proteins, fats, and carbohydrates. A good diet must contain a number of accessory food factors, of which the most important are the vitamins. It must also contain traces of various minerals—compounds of iron, manganese, cobalt, iodine, and so on, as well as ordinary salts. It is worth while noting that vitamins are compounds that occur widely and abundantly in a *natural* diet : were this not so, we could hardly have survived the disability of being unable to synthesize some of them for ourselves. Because vitamins (many of them now synthetic) can be dispensed and taken as pills, the misconception has arisen that man may one day contrive to live on pills of food alone. This idea is wrong, simply because the minimum quantity of even the richest foodstuff that would provide enough energy a day would weigh not less than three-quarters of a pound.

Health and the Individual. It is little use just to look immediately about us if we wish to appreciate what effects our knowledge of bacterial disease and nutrition has had on our personal well-being—for the simple reason that we accept these effects as though they had always been with us. To realize what changes medical science has wrought, we have to contrast our own circumstances with those of starving and disease-ridden areas of the world—for example, parts of India and China and Africa—for the medical advances which impress us most have so far affected only a small part of the surface of the globe. We, for example, are conscious of the fact that modern developments in anaesthesia have made surgery a much less arduous ordeal to bear than it was, say, seventy-five years ago. But anaesthesia is still unknown in many parts of the world. We accept as a commonplace that certain diseased states of the body are due to derangements of the organs which produce hormones, and which help our brain and nerves to integrate the body's various functions. Diabetes, which is due to malfunctioning of the pancreas, is something we all know can be controlled by means of insulin, the hormone produced by certain cells of the pancreas. But insulin is unknown to natives in the heart of Africa or to inhabitants of the plains of Outer Mongolia. The newspapers tell us that crippling arthritic diseases can be held in check by cortisone, a hormone produced by the adrenal gland, and a substance whose molecular structure is sufficiently well known to make one hope that chemists will soon discover how to make it synthetically. But it may be years before cortisone is available outside the countries with well-developed pharmacological industries. And if we do not yet know how to prevent cancer, a disease which is due to cells of the body dividing and spreading in an unrestrained way, or tuberculosis, which is due to a very resistant bacillus, at any rate we appreciate that many scientists are striving to find out how, and that success is not impossible in view of previous achievements of medical science.

MEDICAL PROGRESS AND SOCIETY

Much as we may be tempted to do so, the increase in medical knowledge should not, however, be looked at as an unmixed blessing. Early in the nineteenth century Dr. Malthus, an English clergyman, stirred his contemporaries with the thesis that the unchecked growth of human population could result only in disaster, with teeming millions trying to subsist on too few resources. The growth of industry during the early part of the nineteenth century, with the concomitant increase in the apparently disposable wealth of the world, seemed to have proved him wrong, and up till very recently it was fashionable to suppose that he was only a misguided and dreary pessimist. Such a view is hardly tenable any longer. Increasing control of disease has led to a very considerable fall in the infantile mortality-rate, and also to greater longevity. Because of this and of a concomitant fall in the birth-rate, the age structure of our own society has become violently distorted during the past few decades, since the proportion of people in the reproductive period of life is steadily falling, and in so far as the average age of our population is growing older at an increasingly rapid

rate.[1] What is happening to us in England, and to people in Western
Europe and the United States is, however, exceptional when viewed
against events in the world at large. It is exceptional because, while
progress in medicine has almost everywhere been associated with some
decrease of infantile mortality and in most places with a greater
expectation of life, the net reproductive rate of people in so-called
backward areas has not declined. Economists tell us that birth-rates
in due course adjust themselves inversely to standards of living, in the
sense that the higher the standard of living, the lower the birth-rate.
The fact is, however, that the benefits of medicine spread immeasurably
faster than those of our modern industrial civilization, with the result
that the population of the world is at present growing faster than its
physical resources are being developed. Over the past ten years, it
has increased by 300 millions—that is to say, by a number greater than
the combined total of the populations of the United States of America,
Great Britain, Canada, Australia, and New Zealand. And the bulk of
this increase has occurred in what we refer to as the " backward " areas.
    Let us take a specific example. Until this century, Egypt was a
country of repeated famine. Its food came from the farms of the Nile
Valley, and when, in a season of poor rainfall, the Nile did not overflow
its banks, to water and fertilize the surrounding fields, the crops failed
and the people starved. To prevent this, barrages were built to control
the river, and also to regulate the irrigation of the crops, and a more
efficient agriculture was permitted to flourish. All this economic
improvement went hand in hand with an increased application of
medical knowledge. But in the past fifty years the Egyptian population
has more than doubled, with the paradoxical result that while the river
barrages have removed the fear of crop failures, poverty and hunger
still abound, and in fact may on the average be greater than before.
    Take another example. The virtues of D.D.T. as an insecticide
became widely recognized during the 1939–45 war. Since then D.D.T.
has been increasingly used in various parts of the globe to eliminate

---

[1] Some idea of the mean expectation of life at birth in various countries is given
by the following figures :

India (males) 1931	26·9 years	
Egypt (males) 1927-37	30·2	,,
Ancient Rome (after Macdonell)	25-30	,,
England and Wales (males) 1954	67·6	,,
New Zealand (males) 1934-8	65·5	,,

To-day about 1 in every 6 people in England and Wales is 60 years old or older.
A hundred years ago the proportion was only 1 in 14. It has been estimated that, if
certain current tendencies continue, the proportion may rise to 1 in 4 in 100 years'
time. (See *Reports and Selected Papers of the Statistics Committee* (Royal Commission
on Population), London (H.M.S.O.) 1950).

the mosquito and with it malaria, the principal killing disease of man-kind. It has, for example, been used to great effect since the 1939–45 war both in Cyprus and in British Guiana, two countries where malaria flourished before. To-day Cyprus is said to be clear of the mosquito and British Guiana is almost in the same happy state. But the by-product of this piece of medical progress has been a rapid increase in population. In 1946 the people of British Guiana numbered 370,000, and the number of live births in that year was about 13,400. In 1947 the mosquito was exterminated. By 1949 the number of live births leapt to over 17,000 and to-day the population of the territory is esti-mated to be more than 510,000. Take yet another example—Puerto Rico, a country whose population increased from 1,900,000 to 2,264,000 between 1940 and 1955—a rate of increase out of all proportion to the rate of expansion of the island's resources. And another—Japan, which, at the end of 1947, is estimated to have had a population of 78·5 millions, and which in 1957 had one of over 90 millions.

All experience thus seems to show that public health measures and medical care, however slight, are associated in the less economically developed areas of the world with disproportionate increases in popula-tion. If present trends are any guide, Europeans and people of European descent, who already constitute a small minority of the world's population, will to-morrow be an even smaller one. Neither war nor pestilence can be expected to hold back the vast increase that is occurring in the population of the globe.

Because of this, Julian Huxley has suggested that we may have to take a new view of human destiny. In saying this he is trying to emphasize the possibility that the rate of increase in the population of the globe is outstripping that of the development of its resources. Needless to say, any categorical opinion on this subject is dangerous. History may prove the pessimists to be right, and it is possible that in the end the pressure of population on resources may lead to incalculable chaos. It is, on the other hand, possible that human ingenuity may provide means of developing sources of food supply and of raw materials at a rate sufficient to keep pace with the needs of the growing masses of the world.

Whichever is the better guess, we dare not be blind to the fact that the application of medical knowledge is intimately related to what can be called " population explosions ". World history is thus inevitably bound up with medical progress and practice. Ultimately it will be imperative to combine the application of new biological and medical knowledge with rational population and resources policies. For if the present divorce between the two continues, the benefits which the individual of to-day gains from the advances of medical science will inevitably be paid for by future hardships. Man, we have seen, started off on his human career a million years or so ago. The scientific facts which have so violently transformed his existence are mainly the product of the past fifty or a hundred years. We have time to control the application of our new knowledge, but we must work fast.

LEADING QUESTIONS

BOOK I

THE UNITY OF LIFE

THE UNITY OF LIFE

WHAT is Man? All our deliberations bring us back in the end to this ultimate problem. The question has been asked in all ages and in all countries, and the wide range of answers, reflecting the different outlooks of those speculative thinkers daring enough to attempt to solve the riddle, emphasizes both its importance and its complexity.

Nothing like a complete answer will be attempted in this volume. The approach will be that of the biologist ; we shall look at the problem through the eyes of the scientist who has studied all forms of life, and who regards man as the highest form yet produced by a vast upward-striving process of rivalry called Evolution—to which indeed a thousand ages are but an evening gone.

Admittedly this view is only partial, we might even say narrow, but it is important, for, as Sir Henry Tizard said in his Presidential Address to the British Association in 1948, " Whatever new comforts and luxuries may be provided in the future by the advancement of physical science, it is on the development of the biological sciences that the peace and prosperity of the world will largely depend ".

The biologist makes full use of the information provided by other sciences, and every one of these departments of " organized common-sense " has made its contribution to man's understanding of man, and of his fellow living creatures. It is a peculiar fact, however, that while most of the common technical words of these other sciences have passed into the vocabulary of everyday speech, those of the biologist, and more particularly of that branch of biology which is called " genetics " are only just beginning to be widely known and understood. A glossary of the main scientific terms used in this volume is therefore placed immediately after the final lecture.

Our attention will often be diverted to the lower forms of life, partly because these are of great importance and interest for their own sakes, but also because, as we shall shortly see, there is a strange unity in all forms of life, and we can learn much about the human body by a study of mice or moulds or monkeys.

The whole of this volume is therefore an answer to our first question, and a brief survey of the ground to be covered is given in the following broadcast talk by Dean Matthews. It was one of a series on the nature of man.

WALTER ROBERT MATTHEWS

MAN AS SCIENCE SHOWS HIM

LAST week I tried to open the subject of the nature of man by dwelling on some of the facts which indicate that man is a mystery—perhaps the most mysterious thing in this mysterious universe. Now I hope you will accompany me while I make some reflections on the information which science gives us.

Science, as we understand it, is very modern; it is not much more than two hundred years old. Of course there was science before that time, and men like Aristotle were great pioneers, but the road was not followed with determination and success until the eighteenth century. That was not because our ancestors in the earlier centuries were fools, but because they were interested in other things. Now we are living at the end of a period when the best minds have been concentrated on scientific research, and the changes which have been produced, both in our circumstances and in our ways of thinking, make our civilization quite different from all that preceded it, so that it should be called not Christian civilization but scientific civilization.

What is the distinguishing mark of science? It is the observation of facts. In former times men asked questions enough—indeed all the most fundamental questions were asked long ago in Greece—but they tried to answer them by speculations based on reason. The scientific method is quite different. It does not start by asking tremendous, far-reaching questions, like: What is man? It starts by accumulating facts and it goes on by formulating hypotheses to explain the facts. It is never finished. Its hypotheses are never final; they are always open to revision.

Thus science is always moving on. No one could possibly conceive that a time might come when science was completed and there was no more research to do. Thus we owe to science a new method of obtaining knowledge, a new world, and new light on some of the questions which have been asked ever since men began to think. Naturally science has not neglected the study of man. Some remarkable conclusions have been reached, but I am going to suggest that the mystery of man has not been cleared up. We are left with the old riddle in a new setting.

I suppose the most important change in our conception of man due to science is summed up in the word "evolution". The idea that one species of animal may have developed out of another, the general notion of development, is not new. It had occurred to more than one thinker in earlier centuries. What is new is the testing of this hypothesis by a vast array of facts, and the theory, put forward by Darwin, of natural selection to account for the emergence of new species. What was before a brilliant speculation has become a scientific truth. Now I do not wish to suggest that evolution is a perfectly clear idea. It is not. Nor are the causes of evolution quite certain. There are controversies among experts about the precise value of the natural selection hypothesis, but it does seem to me that the fact of evolution is established beyond question. Man is the product of a process of evolution which makes him more nearly related to the apes than our pride would have led us to suppose.

It would be difficult to exaggerate the effects of this idea of evolution on our whole outlook. We are concerned with the question, What is man? and it would be futile to deny that evolution is very relevant to our answer. Consider the contrast with the view which was generally held only a hundred years ago. It was thought that every species was the result of a special creation by God and that the peculiar dignity of man was secured because he had come into the world by a unique fiat of the Almighty—not the descendant of the lower animals but their lord. It is unfortunate that religious people felt obliged to condemn the theory of evolution when it was first announced, but not surprising, because they had been taught to believe that the Bible contained, not only religious, but scientific truth.

Connected with this theory of evolution is another drastic alteration which science has caused us to make in our thoughts about the human race. The old idea, based on the Bible, was that human beings had been on the earth for only a comparatively short time—or rather that the world itself had not existed for much more than six thousand years. We know now that our species, *homo sapiens*, has immense antiquity, although it is one of the youngest. Of course these calculations are largely based on conjecture, but there is little doubt that beings in all essential respects like ourselves have existed for more than a quarter of a million years. The time during which we have any real history is only a minute fraction of the whole period of human life upon this planet. Of the story of the human race the greater part remains untold.

But not completely untold, for some of the greatest triumphs of science have been in the field of archæology and pre-history. The spade has given us knowledge of men who left no records of themselves. The scientific method of collecting what data there are and forming hypotheses has given us insight into the lives of very remote ancestors. We are able to form an idea of how culture began and developed. We can see the chipped flints which our primitive forefathers used as tools and weapons. We can trace the development from the Stone Age to the great discovery of the use of metals, the Bronze Age and the Iron Age. Though the picture remains obscure in many details, its outline is clear enough. Man, pathetically defenceless compared with other animals, has one great and distinctive quality, reason and inventiveness. He begins to get some kind of control of his environment. And it is interesting to notice that, at a very early stage in his development, he shows that he has the impulse towards art and religion. The famous cave drawings have never been surpassed for vigour. Early burials give indications that man's unconquerable spirit did not despair in the face of death.

We must not forget to mention the new light which has come to us on the ancient civilizations through excavation. We learn that ten thousand years ago there were men living in settled communities, with some kind of government and possessing some knowledge of useful arts. We are able to follow the rise and decay of ancient empires like those of Egypt and Babylon, and to form some idea of the lives and thoughts of the people who lived in them.

The youngest of the sciences—psychology—has, more than any other, directly concerned itself with the question before us, What is man ? It deals with man's mind and his behaviour by the scientific method of observation, experiment, and hypothesis. No doubt psychology is only in its infancy, and consequently much of what it has to say is doubtful and the subject of controversy, but already it has made contributions of the most important kind. Perhaps the most remarkable of its conclusions is the demonstration of the large part which the " unconscious " plays in the life and behaviour of human beings. We are told that the mind of which we are aware is only a small part of the whole mind. Below the level of consciousness is the unconscious, which contains impulses, memories, and tendencies of which we are normally unaware. Very probably in the unconscious there are desires which derive from the animal ancestry of the race and from primitive man. Psychology has shown the decisive part played by our instinctive nature, and that very often when we think we are acting most rationally we are really finding reasonable excuses for doing something by instinct. It has also shown how difficult it is to become a truly unified self, how hard to adapt the instinctive nature, with all its inherited impulses, to the requirements of civilized life.

We have not time to carry this brief sketch of some of the contributions made by science to our problem any further. We must try to sum up their general effect. As I said, the central idea is that of evolution and continuity. The coming into existence of man is the result of a process extending over millions of years. Man's culture and civilization have been developed through thousands of years of struggle with the environment, and, just as his body bears many traces of his kinship with the lower animals, so his mind has elements which belong to more primitive conditions.

The first conclusion we may draw is that man is not at the end of his career. We see him changing and, perhaps, in some respects making progress through the centuries. There is no reason for supposing that the process has now stopped. In fact, it is quite certain that it has not. The human species will not stand still. It must either go forward to new triumphs or degenerate and ultimately disappear. We may conjecture that, if further progress is in store for man, it will be in the sphere of mind and social organization. There is one condition which man has not mastered—his own nature and its impulses. But perhaps speculations about the future of man are unprofitable. Let us turn to conclusions which make no drafts upon the future.

Science presents man to us definitely as a part of nature. In a sense, this is not so much a conclusion as an assumption. We could not study man scientifically at all unless we treated him as if he were an object among other objects. But the assumption has had such brilliant results that it must be partly true. It would be foolish to attempt to deny that we are firmly set in the order of nature. If it is true that we come " trailing clouds of glory " from heaven, it is also true that we come with some highly inconvenient legacies from the earlier stages of our evolution. But it is all the more evident that man is really the crown of the whole of the process. Nothing that we can conclude from science can weaken in the smallest degree the uniqueness of the human race. Man represents an entirely new stage in evolution and his title to that eminence depends, not on his bodily powers, but on his reason.

I must return to a point which I touched upon last week. The mystery of man resides in the fact that he knows he is a mystery. How are we to account for the fact that man, the product of a long process of development from unreasoning things, can seek for truth and can begin to understand the causes which brought him into being ? When we are considering the light which science throws upon our nature we must not leave out the existence of science. One of the most significant things we can say about man is that he has created the science of nature.

The fact of science throws light again upon the real unity of the race. Here we have a body of knowledge which is continually growing by the contributions of men of every race. There is no racial discrimination in science : everyone who will examine the evidence and has the intelligence to understand must agree. There is no appeal to authority or tradition. The truths of science are true in so far as they " stand to reason ". In these days, when the racial prejudices of fanatics are causing untold misery, it is worth while to remember that there is a fellowship of science and of the search for truth which nothing can destroy. In their reason men are one.

You may remember that in my last talk I included, among the problems to be thought about, the suggestion that man seemed to be a splendid failure. I must return to that point. The story of man, as science discloses it, has its elements of heroism. We see the almost defenceless ape-like creature making good through æons of struggle. Finally, we see him, by a unique effort, beginning to understand the world and himself. We exclaim : What a piece of work is man ! And all this is true. But the other side of the picture remains. The power which comes from knowledge has not been used with wisdom. Too often it has been used to hurt and destroy. When we look at the scientific armaments now prepared for the slaughter of nature we get an object lesson on the problem of man. The fact is that the human race, as a whole, is not good enough or wise enough to be entrusted with the forces which it can now control.

Science has much to tell us about the making of man. Among other things it shows us that we are subject to impulses which need to be controlled and redirected. It suggests to us that we may be disastrously divided against ourselves. We may ask whether science can give us the means for controlling and directing its own applications. I do not think it can, because science is concerned with facts and not with values ; it deals with what is and not with what ought to be. The pressing problem is not how to acquire more knowledge —that is going on all the time and will go on whether we like it or not—but how to acquire more wisdom and settled goodwill. I have just said that the next stage in man's development, if he is to advance and not degenerate, would be mental. I must amend it by adding that it must be moral and spiritual if he is not to destroy himself with his own inventions.

These thoughts have necessarily been somewhat disjointed. We have taken a bird's-eye view of a vast stretch of country. I hope you have got the general idea of the long process which went to the making of man and also the idea that man needs remaking if he is to avoid utter failure. To redeem or remake man is the promise of religion.

What have all living things in common ? Almost a century ago, in 1868, T. H. Huxley answered this question in a lecture which quickly became famous—or notorious, according to the point of view. His title was " The Physical Basis of Life ". Only a few years ago Professor J. D. Bernal, a distinguished modern biophysicist, chose the same title for his Guthrie lecture, and in a note to the published paper he said that the positive part of Huxley's lecture " stands as firm to-day as it did when he first put it forward ". Here is this " positive part ", showing not merely a single, but a threefold unity among living things. First there is the unity of function, for all life feeds and grows and reproduces. Then there is the unity of form, for all living things, with one or two exceptions, consist of the same kind of cells. And finally there is the unity of substance, for all forms of life are built of proteins— compounds of carbon, hydrogen, oxygen, and nitrogen.

THOMAS HENRY HUXLEY

THE THREEFOLD UNITY OF LIFE

IN order to make the title of this discourse generally intelligible, I have translated the term " protoplasm ", which is
the scientific name of the substance of which I am about to
speak, by the words " the physical basis of life ". I suppose
that to many the idea that there is such a thing as a physical
basis, or matter, of life may be novel—so widely spread is the
conception of life as a something which works through matter,
but is independent of it ; and even those who are aware that
matter and life are inseparably connected may not be prepared
for the conclusion plainly suggested by the phrase, " *the*
physical basis or matter of life ", that there is some one kind
of matter which is common to all living beings, and that their
endless diversities are bound together by a physical as well
as an ideal unity. In fact, when first apprehended, such a
doctrine as this appears almost shocking to common sense.

What, truly, can seem to be more obviously different from
one another, in faculty, in form, and in substance, than the
various kinds of living beings ? What community of faculty
can there be between the brightly coloured lichen, which so
nearly resembles a mere mineral incrustation of the bare rock
on which it grows, and the painter, to whom it is instinct with
beauty, or the botanist, whom it feeds with knowledge ?

Again, think of the microscopic fungus—a mere infinitesimal ovoid particle, which finds space and duration enough to
multiply into countless millions in the body of a living fly ;
and then of the wealth of foliage, the luxuriance of flower and
fruit which lies between this bald sketch of a plant and the
giant pine of California, towering to the dimensions of a
cathedral spire, or the Indian fig, which covers acres with its
profound shadow, and endures while nations and empires
come and go around its vast circumference. Or, turning to
the other half of the world of life, picture to yourselves the
great Finner whale, hugest of beasts that live, or have lived,

disporting his eighty or ninety feet of bone, muscle, and blubber, with easy roll, among waves in which the stoutest ship that ever left dockyard would founder hopelessly; and contrast him with the invisible animalcules—mere gelatinous specks, multitudes of which could, in fact, dance upon the point of a needle, with the same ease as the angels of the Schoolmen could, in imagination. With these images before your minds, you may well ask what community of form or structure is there between the animalcule and the whale; or between the fungus and the fig tree. And, *a fortiori*, between all four.

Finally, if we regard substance, or material composition, what hidden bond can connect the flower which a girl wears in her hair and the blood which courses through her youthful veins; or what is there in common between the dense and resisting mass of the oak, or the strong fabric of the tortoise, and those broad disks of glassy jelly which may be seen pulsating through the waters of a calm sea, but which drain away to mere films in the hand which raises them out of their element?

Such objections as these must, I think, arise in the mind of everyone who ponders for the first time upon the conception of a single physical basis of life, underlying all the diversities of vital existence; but I propose to demonstrate to you that, notwithstanding these apparent difficulties, a threefold unity —namely, a unity of power or faculty, a unity of form, and a unity of substantial composition—does pervade the whole living world.

No very abstruse argumentation is needed, in the first place, to prove that the powers or faculties of all kinds of living matter, diverse as they may be in degree, are substantially similar in kind.

Goethe has condensed a survey of all the powers of mankind into the well-known epigram:

" Why are the people thus busily moving? For food they are
    seeking,

Children they fain would beget, feeding them well as they
can.
Traveller, mark this well, and, when thou art home, do thou
likewise !
More can no mortal effect, work with what ardour he will."

In physiological language, this means that all the multi-
farious and complicated activities of man are comprehensible
under three categories. Either they are immediately directed
toward the maintenance and development of the body, or they
effect transitory changes in the relative positions of parts of
the body, or they tend toward the continuance of the species.
Even those manifestations of intellect, of feeling, and of will,
which we rightly name the higher faculties, are not excluded
from this classification, inasmuch as to everyone but the subject of them, they are known only as transitory changes in the
relative positions of parts of the body. Speech, gesture, and
every other form of human action are, in the long run, resolv-
able into muscular contraction, and muscular contraction is
but a transitory change in the relative positions of the parts
of a muscle. But the scheme which is large enough to embrace
the activities of the highest form of life covers all those of
the lower creatures. The lowest plant, or animalcule, feeds,
grows, and reproduces its kind. In addition, all animals manifest those transitory changes of form which we class under
irritability and contractility ; and it is more than probable
that, when the vegetable world is thoroughly explored, we
shall find all plants in possession of the same powers, at one
time or other of their existence
I am not now alluding to such phenomena, at once rare and
conspicuous, as those exhibited by the leaflets of the sensitive
plants, or the stamens of the barberry, but to much more
widely spread, and at the same time, more subtle and hidden,
manifestations of vegetable contractility. You are doubtless
aware that the common nettle owes its stinging property to
the innumerable stiff and needle-like, though exquisitely

delicate, hairs which cover its surface. Each stinging-needle tapers from a broad base to a slender summit, which, though rounded at the end, is of such microscopic fineness that it readily penetrates, and breaks off in, the skin. The whole hair consists of a very delicate outer case of wood, closely applied to the inner surface of which is a layer of semi-fluid matter, full of innumerable granules of extreme minuteness. This semi-fluid lining is protoplasm, which thus constitutes a kind of bag, full of a limpid liquid, and roughly corresponding in form with the interior of the hair which it fills. When viewed with a sufficiently high magnifying power, the proto-plasmic layer of the nettle hair is seen to be in a condition of unceasing activity. Local contractions of the whole thickness of its substance pass slowly and gradually from point to point, and give rise to the appearance of progressive waves, just as the bending of successive stalks of corn by a breeze produces the apparent billows of a corn-field.

But, in addition to these movements, and independently of them, the granules are driven, in relatively rapid streams, through channels in the protoplasm which seem to have a considerable amount of persistence. Most commonly, the currents in adjacent parts of the protoplasm take similar directions ; and thus there is a general stream up one side of the hair and down the other. But this does not prevent the existence of partial currents which take different routes ; and sometimes trains of granules may be seen coursing swiftly in opposite directions within a twenty-thousandth of an inch of one another ; while, occasionally, opposite streams come into direct collision, and, after a longer or shorter struggle, one predominates. The cause of these currents seems to lie in contractions of the protoplasm which bounds the channels in which they flow, but which are so minute that the best microscopes show only their effects, and not themselves.

The spectacle afforded by the wonderful energies prisoned within the compass of the microscopic hair of a plant, which we

commonly regard as a merely passive organism, is not easily
forgotten by one who has watched its display, continued hour
after hour, without pause or sign of weakening. The possible
complexity of many other organic forms, seemingly as simple as
the protoplasm of the nettle, dawns upon one; and the comparison of such a protoplasm to a body with an internal circulation,
which has been put forward by an eminent physiologist,
loses much of its startling character. Currents similar to
those of the hairs of the nettle have been observed in a great
multitude of very different plants, and weighty authorities
have suggested that they probably occur, in more or less
perfection, in all young vegetable cells. If such be the case,
the wonderful noonday silence of a tropical forest is, after all,
due only to the dullness of our hearing; and could our ears
catch the murmur of these tiny Maelstroms, as they whirl in
the innumerable myriads of living cells which constitute each
tree, we should be stunned, as with the roar of a great city.

Among the lower plants, it is the rule rather than the
exception, that contractility should be still more openly manifested at some periods of their existence. The protoplasm of
Algæ and *Fungi* becomes, under many circumstances, partially,
or completely, freed from its woody case, and exhibits movements of its whole mass, or is propelled by the contractility
of one, or more, hair-like prolongations of its body, which
are called vibratile cilia. And, so far as the conditions of the
manifestation of the phenomena of contractility have yet been
studied, they are the same for the plant as for the animal.
Heat and electric shocks influence both, and in the same way,
though it may be in different degrees. It is by no means my
intention to suggest that there is no difference in faculty
between the lowest plant and the highest, or between plants
and animals. But the difference between the powers of the
lowest plant, or animal, and those of the highest, is one of
degree, not of kind, and depends, as Milne-Edwards long ago
so well pointed out, upon the extent to which the principle
of the division of labour is carried out in the living economy.

In the lowest organism all parts are competent to perform all
functions, and one and the same portion of protoplasm may
successfully take on the function of feeding, moving, or
reproducing apparatus. In the highest, on the contrary, a
great number of parts combine to perform each function,
each part doing its allotted share of the work with great
accuracy and efficiency, but being useless for any other purpose.

On the other hand, notwithstanding all the fundamental
resemblances which exist between the powers of the protoplasm in plants and in animals, they present a striking difference in the fact that plants can manufacture fresh protoplasm
out of mineral compounds, whereas animals are obliged to
procure it ready made, and hence, in the long run, depend
upon plants. Upon what condition this difference in the
powers of the two great divisions of the world of life depends,
nothing is at present known.

With such qualifications as arise out of the last-mentioned
fact, it may be truly said that the acts of all living things are
fundamentally one. Is any such unity predicable of their
forms ? Let us seek in easily verified facts for a reply to this
question. If a drop of blood be drawn by pricking one's
finger, and viewed with proper precautions, and under a
sufficiently high microscopic power, there will be seen, among
the innumerable multitude of little, circular, discoidal bodies,
or corpuscles, which float in it and give it its colour, a
comparatively small number of colourless corpuscles, of
somewhat larger size and very irregular shape. If the drop
of blood be kept at the temperature of the body, these
colourless corpuscles will be seen to exhibit a marvellous
activity, changing their forms with great rapidity, drawing
in and thrusting out prolongations of their substance, and
creeping about as if they were independent organisms.

The substance which is thus active is a mass of protoplasm, and its activity differs in detail, rather than in principle,
from that of the protoplasm of the nettle. Under sundry
circumstances the corpuscle dies and becomes distended into

a round mass, in the midst of which is seen a smaller spherical body, which existed, but was more or less hidden, in the living corpuscle, and is called its *nucleus*. Corpuscles of essentially similar structure are to be found in the skin, in the lining of the mouth, and scattered through the whole framework of the body. Nay, more ; in the earliest condition of the human organism, in that state in which it has but just become distinguishable from the egg in which it arises, it is nothing but an aggregation of such corpuscles, and every organ of the body was, once, no more than such an aggregation.

Thus a nucleated mass of protoplasm turns out to be what may be termed the structural unit of the human body. As a matter of fact, the body, in its earliest state, is a mere multiple of such units ; and in its perfect condition, it is a multiple of such units, variously modified.

But does the formula which expresses the essential structural character of the highest animal cover all the rest, as the statement of its powers and faculties covered that of all others ? Very nearly. Beast and fowl, reptile and fish, mollusc, worm, and polyp, are all composed of structural units of the same character, namely, masses of protoplasm with a nucleus. There are sundry very low animals, each of which, structurally, is a mere colourless blood-corpuscle, leading an independent life. But, at the very bottom of the animal scale, even this simplicity becomes simplified, and all the phenomena of life are manifested by a particle of protoplasm without a nucleus. Nor are such organisms insignificant by reason of their want of complexity. It is a fair question whether the protoplasm of those simplest forms of life, which people an immense extent of the bottom of the sea, would not outweigh that of all the higher living beings which inhabit the land put together. And in ancient times, no less than at the present day, such living beings as these have been the greatest of rock builders.

What has been said of the animal world is no less true of plants. Imbedded in the protoplasm at the broad, or attached, end of the nettle hair, there lies a spheroidal nucleus. Careful

examination further proves that the whole substance of the
nettle is made up of a repetition of such masses of nucleated
protoplasm, each contained in a wooden case, which is modified
in form, sometimes into a woody fibre, sometimes into a duct
or spiral vessel, sometimes into a pollen grain, or an ovule.
Traced back to its earliest state, the nettle arises as the man
does, in a particle of nucleated protoplasm. And in the
lowest plants, as in the lowest animals, a single mass of such
protoplasm may constitute the whole plant, or the protoplasm may exist without a nucleus.

Under these circumstances it may well be asked, how is
one mass of non-nucleated protoplasm to be distinguished
from another ? why call one " plant " and the other " animal " ?

The only reply is that, so far as form is concerned, plants
and animals are not separable, and that, in many cases, it is
a mere matter of convention whether we call a given organism
an animal or a plant. . . .

Protoplasm, simple or nucleated, is the formal basis of all
life. It is the clay of the potter : which, bake it and paint
it as he will, remains clay, separated by artifice, and not by
nature, from the commonest brick or sun-dried clod.

Thus it becomes clear that all living powers are cognate,
and that all living forms are fundamentally of one character.
The researches of the chemist have revealed a no less striking
uniformity of material composition in living matter.

In perfect strictness, it is true that chemical investigation
can tell us little or nothing, directly, of the composition of
living matter, inasmuch as such matter must needs die in the
act of analysis—and upon this very obvious ground, objections,
which I confess seem to me to be somewhat frivolous, have
been raised to the drawing of any conclusions whatever
respecting the composition of actually living matter, from
that of the dead matter of life, which alone is accessible to us.
But objectors of this class do not seem to reflect that it is also,
in strictness, true that we know nothing about the composition
of any body whatever, as it is. The statement that a crystal of

calc-spar consists of carbonate of lime is quite true, if we
only mean that, by appropriate processes, it may be resolved
into carbonic acid and quicklime. If you pass the same
carbonic acid over the very quicklime thus obtained, you will
obtain carbonate of lime again ; but it will not be calc-spar,
nor anything like it. Can it, therefore, be said that chemical
analysis teaches nothing about the chemical composition of
calc-spar ? Such a statement would be absurd ; but it is
hardly more so than the talk one occasionally hears about the
uselessness of applying the results of chemical analysis to the
living bodies which have yielded them.

One fact, at any rate, is out of reach of such refinements,
and this is, that all the forms of protoplasm which have yet
been examined contain the four elements, carbon, hydrogen,
oxygen, and nitrogen, in very complex union, and that they
behave similarly towards several reagents. To this complex
combination, the nature of which has never been determined
with exactness, the name of Protein has been applied. And
if we use this term with such caution as may properly arise
out of our comparative ignorance of the things for which it
stands, it may be truly said, that all protoplasm is protein-
aceous, or, as the white, or albumen, of an egg is one of the
commonest examples of a nearly pure protein matter, we may
say that all living matter is more or less albuminoid. . . .

And now, what is the ultimate fate, and what the origin,
of the matter of life ?

Is it, as some of the older naturalists supposed, diffused
throughout the universe in molecules, which are indestructible
and unchangeable in themselves ; but, in endless transmigra-
tion, unite in innumerable permutations into the diversified
forms of life we know ? Or is the matter of life composed of
ordinary matter, differing from it only in the manner in which
its atoms are aggregated ? Is it built up of ordinary matter,
and again resolved into ordinary matter when its work is done ?

Modern science does not hesitate a moment between these
alternatives. Physiology writes over the portals of life,

" Debemur morti nos nostraque " [We and ours are destined to die] with a profounder meaning than the Roman poet attached to that melancholy line. Under whatever disguise it takes refuge, whether fungus or oak, worm or man, the living protoplasm not only ultimately dies and is resolved into its mineral and lifeless constituents, but is always dying, and, strange as the paradox may sound, could not live unless it died.

In the wonderful story of *La Peau de chagrin*, the hero becomes possessed of a magical wild ass's skin which yields him the means of gratifying all his wishes. But its surface represents the duration of the proprietor's life ; and for every satisfied desire the skin shrinks in proportion to the intensity of fruition, until at length life at the last handbreadth of the *peau de chagrin* disappears with the gratification of a last wish.

Balzac's studies had led him over a wide range of thought and speculation, and his shadowing forth of physiological truth in this strange story may have been intentional. At any rate, the matter of life is a veritable *peau de chagrin*, and for every vital act it is somewhat the smaller. All work implies waste, and the work of life results, directly or indirectly, in the waste of protoplasm.

Every word uttered by a speaker costs him some physical loss ; and in the strictest sense, he burns that others may have light—so much eloquence, so much of his body resolved into carbonic acid, water, and urea. It is clear that this process of expenditure cannot go on for ever. But, happily, the protoplasmic *peau de chagrin* differs from Balzac's in its capacity of being repaired, and brought back to its full size, after every exertion.

For example, this present lecture, whatever its intellectual worth to you, has a certain physical value to me, which is, conceivably, expressible by the number of grains of protoplasm and other bodily substance wasted in maintaining my vital processes during its delivery. My *peau de chagrin* will be distinctly smaller at the end of the discourse than it was at the beginning. By and by, I shall probably have recourse to the

substance commonly called mutton, for the purpose of stretch-
ing it back to its original size. Now this mutton was once the
living protoplasm, more or less modified, of another animal—
a sheep. As I shall eat it, it is the same matter altered, not
only by death, but by exposure to sundry artificial operations
in the process of cooking.

But these changes, whatever be their extent, have not
rendered it incompetent to resume its old functions as a matter
of life. A singular inward laboratory, which I possess, will
dissolve a certain portion of the modified protoplasm; the
solution so formed will pass into my veins; and the subtle
influences to which it will then be subjected will convert the
dead protoplasm into living protoplasm, and transubstantiate
sheep into man.

Nor is this all. If digestion were a thing to be trifled with,
I might sup upon lobster, and the matter of life of the crustacean
would undergo the same wonderful metamorphosis into
humanity. And were I to return to my own place by sea, and
undergo shipwreck, the crustacea might, and probably would,
return the compliment, and demonstrate our common nature
by turning my protoplasm into living lobster. Or if nothing
better were to be had, I might supply my wants with mere
bread, and I should find the protoplasm of the wheat plant
to be convertible into man, with no more trouble than that of
the sheep, and with far less, I fancy, than that of the lobster.

Hence, it appears to be a matter of no great moment what
animal or what plant I lay under contribution for protoplasm,
and the fact speaks volumes for the general identity of that
substance in all living things. I share this catholicity of assimilation with other animals, all of which, so far as we know,
could thrive equally well on the protoplasm of any of their
fellows, or of any plant; but here the assimilative powers of
the animal world cease. A solution of smelling-salts in water,
with an infinitesimal proportion of some other saline matters,
contains all the elementary bodies which enter into the composition of protoplasm; but, as I need hardly say, a hogshead

of that fluid would not keep a hungry man from starving, nor
would it save any animal whatever from a like fate. An animal
cannot make protoplasm, but must take it ready-made from
some other animal or some plant—the animal's highest feat
of constructive chemistry being to convert dead protoplasm
into that living matter of life which is appropriate to itself.

Therefore, in seeking for the origin of protoplasm, we must
eventually turn to the vegetable world. The fluid containing
carbonic acid, water, and ammonia, which offers such a
Barmecide feast to the animal, is a table richly spread to
multitudes of plants ; and, with a due supply of only such
materials, many a plant will not only maintain itself in vigour,
but grow and multiply until it has increased millionfold, or
a million millionfold, the quantity of protoplasm which it
originally possessed ; in this way building up the matter of
life to an indefinite extent from the common matter of the
universe.

Thus, the animal can only raise the complex substance of
dead protoplsam to the higher power, as one may say, of
living protoplasm ; while the plant can raise the less complex
substances—carbonic acid, water, and nitrogenous salts—to
the same stage of living protoplasm, if not to the same level.
But the plant also has its limitations. Some of the fungi, for
example, appear to need higher compounds to start with ;
and no known plant can live upon the uncompounded elements
of protoplasm. A plant supplied with pure carbon, hydrogen,
oxygen, and nitrogen, phosphorus, sulphur, and the like,
would as infallibly die as the animal in his bath of smelling-
salts, though it would be surrounded by all the constituents
of protoplasm. Nor, indeed, need the process of simplification of vegetable food be carried so far as this, in order to
arrive at the limit of the plant's thaumaturgy. Let water,
carbonic acid, and all the other needful constituents be
supplied except nitrogenous salts, and an ordinary plant will
be unable to manufacture protoplasm.

Thus the matter of life, so far as we know it (and we have

no right to speculate on any other), breaks up, in consequence of that continual death which is the condition of its manifesting vitality, into carbonic acid, water, and nitrogenous compounds, which certainly possess no properties but those of ordinary matter. And out of these same forms of ordinary matter, and from none which are simpler, the vegetable world builds up all the protoplasm which keeps the animal world a-going. Plants are the accumulators of the power which animals distribute and disperse.

It is certainly surprising, as Professor Bernal pointed out, that Huxley did not include a fourth " unity "—that of a common history—in his lecture. It seems to be a well-established fact that all forms of life to-day arise from pre-existing forms. No new life is created from non-living material, although, until the work of Pasteur finally refuted the theory of spontaneous generation, it was widely believed that the simpler forms might so arise. Even the credulity of van Helmont, who thought that mice could be brought into existence by the simple method of leaving some dirty linen and a few grains of wheat in a cupboard, was possibly still common even in Pasteur's time among the ignorant peasants. *Does life still arise from non-living material?* Here are some extracts from Pasteur's *Memoir on the Organized Corpuscles which exist in the Atmosphere*. The careful exploration of every possibility which he showed in these experiments was typical of one of the greatest of all benefactors of mankind.

LOUIS PASTEUR

THE DOCTRINE OF SPONTANEOUS GENERATION

IN ancient times, and until the end of the Middle Ages, everyone believed in the occurrence of spontaneous generation. Aristotle says that animals are engendered by all dry things that become moist and all moist things that become dry.

Even in the seventeenth century, many authors give methods for producing frogs from the mud of marshes, or eels from river water.

Such errors could not survive for long the spirit of investigation which arose in Europe in the sixteenth and seventeenth centuries.

Redi, a celebrated member of the *Accademia del Cimento*, demonstrated that the worms in putrefying flesh were larvæ from the eggs of flies. His proofs were as simple as they were decisive, for he showed that surrounding the putrefying flesh with fine gauze absolutely prevented the appearance of these larvæ. . . .

But, in the second part of the seventeenth and the first part of the eighteenth centuries, microscopic observations rapidly increased in number. The doctrine of spontaneous generation then reappeared. Some, unable to explain the origin of the varied organisms which the microscope showed in their infusions of animal or vegetable matters, and seeing nothing among them which resembled sexual reproduction, were obliged to assume that matter which has once lived keeps, after its death, a special vital force, under the influence of which its scattered particles unite themselves afresh under certain favourable conditions with varieties of structure determined by these conditions.

Others, on the contrary, used their imagination to extend the marvellous revelations of the microscope, and believed they saw males, females, and eggs among these Infusoria, and they consequently set themselves up as open adversaries of spontaneous generation.

One must recognize that the proofs in support of either of these opinions scarcely bore examination.

[Pasteur was led to a study of this problem because of his work on the fermentation of beer, wine, milk, etc. He wanted to prove his theory that all ferments are living organisms like yeast, and his line of attack was to show that they can only

arise from their own germs or spores ; they were not produced
by the action of air on albuminous substances, as most people
thought.]

My first care was to find a method which should permit of
collecting the solid particles that float in the air and studying
them under the microscope. It was first necessary to remove
if possible the objections held by the partisans of spontaneous
generation against the ancient hypothesis of the aerial dissemination of germs. . . .

The method which I used to collect and examine the dusts
suspended in the air is very simple ; it consists in filtering a
known volume of air through gun-cotton, which is soluble in
a mixture of alcohol and ether. The solid particles collect
on the fibres of the cotton. The cotton is then treated with
its solvent, and after a time all the solid particles fall to the
bottom of the liquid ; they are washed several times and
transferred to the stage of the microscope, where they are
easily examined. . . .

These simple manipulations allow one to observe that
ordinary air always contains a variable number of corpuscles
whose form and structure show them to be of organic nature.
In size they vary from the smallest diameters up to 0·01 or
0·015 of a millimetre, or more. Some are perfectly spherical,
others oval. Their outlines are more or less sharply defined.
Some are quite transparent, but others have a granular substance and are opaque. Those which are transparent with
clearly defined outlines are so much like the spores of common
moulds that the cleverest microscopist could not distinguish
between them. That is all that one can say, just as one can
only affirm that among the others there are some which
resemble encysted Infusoria or the globules which are regarded
as the eggs of these minute creatures. . . .

I found that a little wad of cotton, thus exposed for twenty-
four hours in the summer after a spell of fine weather to a
current of one litre of air a minute from the rue d'Ulm,

collects several thousands of organized corpuscles. The number varies indefinitely with the state of the atmosphere—before or after rain, in still or windy weather, by day or by night, near the ground or at some distance from it. . . .

As we have just seen, organized corpuscles are always to be found suspended in the air ; these cannot be distinguished from the germs of inferior organisms by their shape, size, or apparent structure, and are present in numbers that, without exaggeration, are indeed great. Do fertile germs really exist among them ? Obviously this was the interesting question ; I believe I have found a definite answer. But before describing the experiments which bear more particularly on this side of the subject, it is necessary to consider whether Dr. Schwann's observations on the inactivity of air which has been heated are well established. . . .

[Pasteur boiled a solution of sugar in water with a little brewer's yeast for two or three minutes. Then it was allowed to cool, the flask filling with air which was drawn into it over red-hot platinum. The flask was then sealed, and, no matter how long it was kept, the solution remained clear, without any trace of organisms or of fermentation in it. If the flask was allowed to fill with ordinary air, a day or two sufficed to show the growth of organisms.]

The results of these experiments have taught us :

(1) that in suspension in ordinary air there are always organized corpuscles which closely resemble the germs of inferior organisms ;

(2) that sugared yeast-water, though eminently alterable when exposed to ordinary air, remains unaltered, limpid, without producing Infusoria or moulds, when left in contact with air that has been previously heated.

This admitted, let us endeavour to find out what will happen if into this albumen-containing sugar solution are sown the dusts of which the collection is described above, without introducing anything else but the dusts, and only air that has been heated. . . .

Here are the details of (one) experiment.

Early in November 1859 I prepared several flasks of 250 cubic centimetres capacity, containing 100 cubic centimetres of sugared yeast-water and 150 cubic centimetres of heated air. They remained in the incubator at a temperature of about 30° till the 8th of January 1860. On that day, about 9 A.M., I introduced into one of these flasks a portion of a wad of cotton loaded with dusts, collected as I described in chapter II.

On the 9th of January, at 9 A.M., nothing particular could be seen in the liquid in the flask. On the same day at 6 P.M., one could see very distinctly little tufts of mould growing out from the tube with the dusts. The liquid was perfectly clear.

On the 10th of January at 5 P.M., besides the silky tufts of the mould, I saw on the walls of the flask a large number of white streaks which looked iridescent on holding the flask between the eye and the light.

On the 11th of January, the liquid had lost its clearness. It was all turbid, to such an extent that one could no longer see the tufts of mould.

I then opened the flask by a scratch of the file and studied under the microscope the different growths which had appeared.

The turbidity of the liquid was due to a crowd of little bacteria of the smallest dimensions, very rapid in their movements, spinning sharply or swaying to and fro, etc.

The silky tufts were formed by a mycelium of branching tubes.

Finally, the precipitate which showed itself on the 10th of January in the form of white streaks was composed of a very delicate Torula . . . resembling brewer's yeast, but with smaller cells. . . .

Here then were three growths derived from the dusts which had been added, growths of the same kind as those which appear in similar sugared albuminous liquids when they are abandoned to the contact of ordinary air. . . .

I could multiply many times such examples of growths in sugared yeast-water following on the addition of dusts from air, in an atmosphere of air previously heated and of itself quite inactive. I have chosen for description an experiment which showed very common organisms, which occur frequently in such liquids as those which I employed. But the most diverse organisms may appear. . . .

One might wonder if, in the preceding experiments, the cotton, as an organic substance, had some influence on the results. It would above all be useful to know what would happen if similar manipulations were carried out on flasks prepared as described, without the atmospheric dusts. In other words, has the method of introducing the dusts any influence of its own ? It is indispensable to know this.

In order to answer these questions, I replaced the cotton by asbestos. Little wads of asbestos, through which a current of air had been passed for several hours, were introduced into the flasks according to the preceding instructions, and they gave results of exactly the same kind as those we have just quoted. But with wads of asbestos previously calcined and not filled with dust, or filled with dust but heated afterwards, no turbidity, nor Infusoria, nor plants of any kind were ever produced. The liquids remained perfectly clear. I have repeated these comparative experiments very many times, and I have always been surprised by their distinctness, by their perfect constancy. It would seem, indeed, that experiments of this delicacy should sometimes show contradictory results due to accidental causes of error. But never once did any of my blank experiments show any growth, just as the sowing of dusts has always furnished living organisms.

In face of such results, confirmed and enlarged by those of the following chapters, I regard this as mathematically demonstrated : all organisms which appear in sugared albuminous solutions boiled and then exposed to ordinary air derive their origin from the solid particles which are suspended in the atmosphere.

But, on the other hand, we have seen in chapter II that these solid particles include, amid a multitude of amorphous fragments—carbonate of lime, silica, soot, bits of wool, etc., organized corpuscles which are so like as to be indistinguishable from the little spores of the growths whose formation we have recognized in this liquid. These corpuscles are then the fertile germs of the growths.

We may conclude, moreover, that if an albuminous solution of sugar in contact with air which has been heated does not alter, as Dr. Schwann first observed, it is because the heat destroyed the germs which the air was carrying. All the adversaries of spontaneous generation had foreseen this. I have done nothing but supply sure and decisive proofs, obliging non-prejudiced minds to reject all idea of the existence in the atmosphere of a more or less mysterious principle, gas, fluid, ozone, etc., having the property of arousing life in infusions.

If no new forms of life are now coming into existence, it would seem to be a fair assumption that this has long been the case, and that life arose on this planet long ages ago under conditions which no longer hold. *How did life begin?* Professor Bernal devoted the greater part of his lecture on " The Physical Basis of Life " to the answering of this question—after first pointing out the immense scope and difficulty of the problem. It used to be said, with an easy nonchalance, that life arose in the warm seas of the primeval world. Greater knowledge, as is so often the case with scientific theories, has shown that while this first assumption may be fundamentally correct, it glosses over a whole host of technical difficulties. Professor Bernal dealt with some of these in his lecture. He traced three critical stages. The first was the formation of large chemical molecules which, by virtue of their complexity and the store of energy they contained, allowed for the possibility of living processes, that is, in J. B. S. Haldane's words, for a " self-perpetuating pattern of chemical reactions ". The second stage was the liberation of such potentially living molecules from reliance on their own stores of energy, by a change over to photosynthesis, that is, to a dependence on the energy of sunlight. The third was the emergence among such large molecules, or rather accumulations of molecules, of the cell, and thus of specific organisms. Here is an extract from this important lecture.

JOHN DESMOND BERNAL

HOW LIFE BEGAN

BY the time the earth had cooled sufficiently for the water to condense, the atmosphere may have been largely one of nitrogen with a gradually decreasing concentration of carbon dioxide. . . . The seas would contain primarily ammonia, carbon dioxide, and hydrogen sulphide in solution. Whether they also contained salt in anything like the present concentration is still an open question, though there are enough minerals containing chloride to provide for present concentration if sufficient rock had been worked over in the course of geological history by weathering processes ; but, on the other hand, it is possible that vapourized halides formed parts of the original atmosphere, came down molten, solidified, and were dissolved in the primitive seas.

The surface of the world at this stage must not have been very different from what it is now, except for the bareness of the rocks and consequently greater speed of weathering. We still do not understand the mechanism of continent and mountain formation, but there is no reason to believe that it was dependent on organic processes. We may therefore assume most of the geographical features which we now observe, with the exception of coral islands, though inorganic calcium carbonate precipitation may have occurred. In particular there must have been, as now, extensive areas of mud in deltas and on continental shelves, some of which would be exposed to the tides. Correspondingly, on the land a kind of soil would be formed wherever the run-off was not sufficient to remove weathered material or where it was deposited in rivers, and this soil would also contain clay.

At first there might be no reason why the world should not have continued in this state indefinitely, but there was an active agent operating at that time which is no longer in operation, namely the influx of solar radiation out to the far ultra-violet of 2000 Å [1] or less. If we put to a physical chemist the problem of the reactions occurring in a weak solution of ammonium carbonate and sulphide under such radiation, he would agree that, although it was not possible without experiment to determine exactly what compounds would be formed, a process of polymerization and condensation leading to the formation of nitrogenous organic compounds such as the amino-acids is almost certain, and would proceed until an equilibrium was reached where breakdown was equal to formation. It would be of the utmost importance to make these vague remarks more precise, and researches in ultra-violet photosynthesis would certainly be of interest, and might be of practical use.

Equilibrium, moreover, could be reached in two ways. One would be the straight photochemical process of absorption and emission ; the other might well be a dark breakdown through a number of intermediate compounds liberating energies corresponding to smaller quantum steps and longer wavelengths. Life from the purely physico-chemical point of view is simply a denotation of the complex mechanism of this latter process. . . .

The stage is now set for the appearance of life itself. In Moleschott's classic phrase, " It is woven out of air by light ". It is here that organic chemistry first begins. Condensations and dehydrogenations are bound to lead to increasingly unsaturated substances, and ultimately to simple and possibly even to condensed ring structures, almost certainly containing nitrogen, such as the pyrimidines and purines. The appearance of such molecules makes possible still further syntheses.

[1] Å : Ångstrom unit = one ten-millionth of a millimetre.

The primary difficulty, however, of imagining processes going thus far is the extreme dilution of the system if it is supposed to take place in the free ocean. The concentration of products is an absolute necessity for any further evolution. One method of concentration would of course take place in lagoons and pools which are bound to have fringed all early coastlines, produced by the same physical factors of wind and wave that produce them to-day. It has occurred to me, however, that a much more favourable condition for concentration, and one which must certainly have taken place on a very large scale, is that of adsorption in fine clay deposits, marine and fresh-water. Our recent knowledge of the structures of clays has shown what an enormous rôle they still play in living processes. There is probably to-day more living matter, that is protein, in the soil and in the estuarine and sea-bed clays than above the surface or in the waters. Now the effective part of this fine-grained clay is known, particularly through the electron microscope studies of Hast (1947), to consist of what might reasonably be called clay molecules, single layers of aluminium silicate some ten Ångstroms thick and a hundred and forty across, covered on both surfaces with hydroxyl groups and capable of adhering with a larger or smaller number of water molecules into small pseudo-crystals like piles of coins. Such a small clay particle has an enormous effective adsorptive surface. It has already been shown, particularly by McEwen (1948), that organic chemicals of a wide variety are pre-ferentially adsorbed on such surfaces in a regular way. It is therefore certain that the primary photochemical products would be so adsorbed, and during the movement of the clay might easily be held blocked from further possibly destructive transformations. In this way relatively large concentrations of molecules could be formed.

This formation is impossible nowadays for two reasons : firstly the cutting off of the ultra-violet by the ozone layer, and secondly the almost universal presence of life, which would destroy such molecules if they were formed in any other way. The very absence of life ensures an accumulation of material containing available energy for indefinite periods : the original world was sterile. Now the clay molecules have another property besides adsorption. Small molecules attached to them are not fixed at random, but in definite positions relative not only to the clay but to each other, and, held in such positions, they can interact and form more complex compounds, especially if energy can be supplied in the form of light. Clays are now one of the most important of industrial catalysts. Polymerization would take place particularly easily with unsaturated compounds, with relatively free electrons. It is in this way that we may imagine that simpler molecular compounds could be made to undergo complex polymerization, polymerization to such an extent that the macromolecules produced might be able to persist in a colloidal form even without clay, and become catalysts or, as we should now call them, enzymes in their turn.

Clay is not the only material on which adsorption may take place. Quartz is another very active material which, as sand, would occur separately or together with clay. The importance of quartz in the formation of primitive molecules out of which life is constructed may be a crucial one in that quartz is the only common mineral possessing asymmetrical structure, some crystals having a right-handed twist and others a left. The characteristic of molecules occurring in living organisms, first brought out by Pasteur a hundred years ago, is that they are also asymmetric, and it has always been a very great difficulty to explain life originating with such molecules, as normal chemical processes produce right- and left-handed molecules with equal facility. It may be, of course, as Pasteur himself thought, that some general feature in the environment favours one rather than the other type, for example, the rotary polarization of moonlight or the magnetic moment of the earth ; but to me it seems more plausible that the particular twist was given at one time by the preferential adsorption of a pair of asymmetric molecules on quartz, and, as Mills has shown, once one asymmetric isomer was produced, even locally, it would produce a situation in which ultimately only one kind could be formed.

So far we have followed the track from the inorganic world by steps which, though they cannot be indicated in detail for the lack of necessary research, most of which is quite feasible, nevertheless are not only plausible, but, with our present knowledge, inevitable. However, to get any significant information we must look at the other end of the play and try to draw deductions from the inner chemical structure of actual organisms. Now here we are met at the outset with a remarkable set of facts, which call for explanation in terms of origin. The overwhelming majority of living organisms, from the lowest bacteria to trees and men, are all built of a relatively small number (about thirty in all) of types of chemical molecules containing between four and forty atoms in each. Every chemical molecule has its origin in some previous combination, as certainly as every atom of every element. More complex forms, particularly in this or that organism, can be derived from those simple types, and the few cases which have been studied have been shown to be so derived.

From this it follows that there is only *one* predominating life, derived from one common chemical basis. This is exactly the same logic that shows, for example, that all so-called Indo-European languages have a common set of root words, however much they have deviated afterwards. It would be wrong, however, to assume that the common molecules, the amino-acids, the sugars and the purines which are the joint stock of existing life are necessarily the first chemicals, because there are to be found some aberrant bacteria : the purple and green sulphur bacteria, which do not contain some of these mole-cules, in particular free sugar. It may be that modern life, as we may call it, represents a second stage, and we may have to reconstruct the first stage from these particularly primitive survivals. But all life, including these, contains one group of compounds of a far more complex nature which does seem to be of crucial importance, namely, the proteins.

A hundred years ago Engels referred to life as the " mode of action of albuminous or protein substances ", and bio-chemical advances have only confirmed this dictum and made it more precise. The work of the last fifty years has shown something both of their function and of their structure. So many of the chemical reactions occurring in living systems have been shown to be catalytic processes occurring isotherm-ally on the surface of specific proteins, referred to as enzymes, that it seems fairly safe to assume that all are of this nature and that the proteins are the necessary basis for carrying out the processes that we call life. Now, although since the great work of Fischer we know that the proteins consist of various combinations of some twenty amino-acids, we still do not know the precise structure of any of them. But we do know that they have a precise structure, and we have reasonable hope of determining it in the not so very distant future. It is perhaps significant that, though the number of different proteins may be counted in tens of millions, this represents an insignificant proportion of the possible combinations of twenty amino acids. The most likely explanation is that certain sub-units containing the same amino acids in the same order must occur over and over again.

The work of the physico-chemists, particularly of Sved-berg, has shown that active proteins exist in the form of molecules of definite molecular weight, and more recently X-ray structure analysis has shown that they are perfectly definite chemical compounds with identical molecules which persist unchanged through various grades of crystal hydra-tion and into solution. We possess already much information as to the actual molecular arrangement, but unfortunately its full interpretation is a task for the future. The situation at the moment is extremely similar to that of an archæological expedition that has discovered large quantities of rock inscrip-tions in an unreadable script. They may not know what the inscriptions mean, but they do know they mean something, and they may reasonably hope to decipher them.

The pattern of the inorganic molecule—common salt or water, for example—is like the decimal we get from a fraction like $\frac{1}{2}$ or $\frac{1}{4}$. One or two figures, 0·5 or 0·25, and the story is completely told. But carbon atoms can link together in long chains, and so the organic molecules, all of which contain carbon, are not limited in size. Their symbol is more like the recurring decimals we get from fractions such as $\frac{1}{3}$ or $\frac{1}{11}$—0·33 recurring, or 0·0909 for as long as we like. And on the same analogy the pattern of the first living molecule would then be represented by the decimal we get from a large prime number, in which a long string of figures recurs again and again in that " self-perpetuating pattern " of which Haldane spoke. Not only this, but when such a decimal is multiplied by 2 or 3 or more, we still have the same pattern— the very same figures in the very same order, although the sequence naturally begins in a different place. One-seventh, to take the simplest example, is 0·142857 recurring, while $\frac{2}{7}$ is 0·285714 recurring.

The third great step was the formation of living cells, and, with one or two exceptions like the slime-moulds, all living things are composed of cells. These cells are much more complicated organisms than was realized until quite recently, and although the full story would now fill several very large volumes we are still far from knowing all about them. A number of different sciences have made their contributions. *What can the chemist tell us about the living cell?* Dr. Philip Eggleton, formerly Reader in Biochemistry at Edinburgh, outlines the main facts.

PHILIP EGGLETON

THE CHEMISTRY OF LIFE

A HUMAN body—yours, say, or mine—is built of chemical atoms, about a thousand million million million million of them, and it is the avowed object of biochemistry to find out what they are all doing there.

Such an ambitious project would be sheer lunacy if it were not for two things which simplify matters. The first is that atoms practically never exist alone. We do not find atoms existing as atoms. Rather, they tend to unite into groups with very definite patterns, groups or structures which the chemist calls molecules. When a lot of molecules of the same kind are brought together they tend to arrange themselves in still larger patterns called crystals. Thus it comes about that when a chemist gets crystals forming out of some solution he can usually be sure that all the molecules that make up each crystal are of the same kind.

So one part of the immensely difficult problem of trying to find out about the chemistry of life resolves itself into the question, how many different kinds of pure substance or crystal can the chemist isolate from the tissues of a plant or animal? For this tells us about the molecular patterns that are being used in their bodily architecture. The answer is surprising. More than 99 per cent. of our bodies is accounted for by about three dozen such crystalline materials. Only three dozen—out of the hundreds of thousands that chemists have themselves learned to make in the last hundred years. They are not even out-of-the-way rarities, most of them : water, common salt, glucose—these three at least can be found in most kitchens.

The second consideration that eases the path of the biochemist is the fact that it is the patterns of the atom-groupings that matter, not the atoms themselves. This is clear at once if you remember that your body structure is constantly wearing out and being replaced from your food. More than half the atoms in your body to-day were not there a week ago, and will be gone next week. Hardly anything at all in your body now was there five years ago. Yet you rightly consider you are still the same person. Why? Because the patterns in your architecture are still the same. Every bit has been renewed, but the new atoms have been fitted into the old patterns. So the job of the biochemist is reduced first to finding the different molecular patterns that are used in our bodies, and then to finding how these are fitted together to make the living cells, which in turn make up your body.

I said that 99 per cent. or so of your body—or of any part of your body—is accounted for by some three dozen simple crystalline compounds. That much was known many years ago and I am not going to say any more about them. The interest of biochemists to-day centres on that 1 per cent. or so that is not thus accounted for. Shall we find something mysterious there? A little touch of something that makes all the difference between the living and the dead? We have not so far, and we do not seriously expect to. There is a great deal that is mysterious about a living body, but not at this level. Difficulties we expect, increasing difficulty, but only because the easy things were naturally done first, and the chemical molecules we are now trying to get out and purify are often present only in tiny traces. A few years ago the Dutchman Kogl had to start with several tons of egg yolk in order to isolate a few milligrams, about a pin's head, of a certain crystalline substance he was after. And that was not exceptional. Looking for needles in haystacks is child's play by comparison. You can always bring a magnet to bear in your search for the missing needle, but there is no such short cut for the biochemist.

What are these substances the biochemist is now hunting, these compounds present only in traces, and making up the last 1 per cent. or so of the list? They are all of first-class importance in spite of their rarity and they go by such names as vitamins, hormones and co-enzymes. Everybody has heard of vitamins. I am not going to extol their virtues : I just want to point out that a vitamin is any compound needed in traces by our body machinery, but which our bodies cannot make for themselves. Needed in traces, mark you ; that is why they were hard to spot and difficult to extract and purify. Our bodies become diseased if we do not get enough of the various vitamins. This has consequences of great interest to physicians and even to politicians, but the biochemist is interested in another way. He wants to know what the vitamins are doing in our bodies that their absence should be so disastrous. Surely they must be doing something important if the lack of a trace is fatal. There is no general answer, but it seems that vitamins are promoters of definite chemical processes in the body. Now the biochemist finds among the materials in any tissue, dozens of other promoters of chemical reactions, but our bodies are able to make their own supply of most of them. It seems to me possible that the vitamins are the odd few of those promoter substances that we cannot make for ourselves.

The accepted name for many of these promoter substances is " co-enzymes ", and the point I want to make about them is that they are not, in general, complicated substances. They can be crystallized and some have already been made in the laboratory by the organic chemists, and most of them will be, I have no doubt, in the not too distant future.

Hormones are another set of materials that interest the biochemist a lot to-day. A hormone is a chemical compound —often a fairly simple one—which is known to be manufactured in the body in special organs called glands. These hormones travel all over the body, dissolved in the blood, and produce the most remarkable effects. One of the first to be studied—adrenaline—is released into the blood in emergencies, and wherever it goes it promotes changes of various kinds, but all are such as to prepare us for violent exercise. Sugar appears in the blood ready to be used as fuel. The heart starts to beat harder and faster. The pupils dilate. Unessential organs cut down their blood supply to spare blood for the muscles. And so on. And all done by a small trace of a simple chemical, adrenaline. We know of a dozen or so other hormones in our bodies, and no doubt there are more to come ; but not many more. You see, the glands that produce them are all easily recognizable and they have all been studied with some care. Only one gland—the thymus—

still has us seriously guessing. The thymus gland is in the
chest, and it seems to be important to a baby, to judge by its
size, but not important to adults. What chemical substance
it produces, we still do not know.

But speaking of babies, everyone, I imagine, has at some
time wondered how a hen's egg manages to turn into a chick.
It is the sort of thing that would be classed as a miracle if
we had not seen it happen so often. How does this almost
formless mass of food inside the shell organize itself into the
intricate living pattern that we call the chick? This is a
thrilling problem, if you have a taste for problems, as most of
us have. I do not see that the answer, when we get it, will be
any use to us, but it will be enormously interesting to work it
out. A beginning has been made already, though it is easier
to work with sea urchins' eggs than with hens'. If you cut a
sea urchin embryo in two, and do it clearly enough, each of
the two halves develops into a complete sea urchin. Clearly
the cells that make up the embryo must be at that stage pretty
versatile. But later on you cannot do this any more. One part
of the embryo has begun to take charge of all the rest, and
now any bit of the rest that is cut off has lost its leader, and
it dies. How does the leader-bit control the rest? Remember
there are as yet no nerves, no glands, no organs of any kind.
That the leader-bit does control the rest is proved by cutting
it out of one embryo and inserting it in another, whereupon
it takes charge of the development in its neighbourhood,
tending to produce a sort of siamese twin, because the second
embryo already had a leader of its own, and it now has two.

It was a pretty obvious guess that the leader tissue must be
making some chemical substance which is diffusing into the
rest of the embryo and affecting its growth. If this were so
it should be possible to extract this chemical from bits of
leader tissue and find out what it is. And indeed this deduction
is being put to the test now. The active chemical is present
in the juice squeezed out of such tissue, for this juice can
produce effects of the expected kind on the growth of embryos.

But the rest of the job has still to be done. Of course, even
when this problem has been cleared up, we still have not
explained why an egg turns into a living creature, but we shall
know something of *how* it happens.

But now let us stop a minute and consider. Our bodies are
mechanisms built out of a small number of crystalline substances, and in addition there are traces present of other
things—hormones, co-enzymes, and so on, most of them also
fairly simple substances—which are needed in running the
machine. The total number of these may eventually prove
to be very large, but it does not seem likely. Does all this
sound to you too simple ? Can a human being with all his
potentialities really be described in this way ? Before you
make up your mind, remember that there are only fifty-two
playing cards in a pack, yet the number of different hands that
can be dealt is so enormous that I doubt if any card player has
ever been dealt out the same hand twice. The apparently
small number of simple chemicals used in the structure and
running of our bodies is quite consistent with the most elaborate architecture and the most complex behaviour.

Nowhere is this paradox sharper than in the mysterious
question of inheritance. Why does a human being take after
his parents ? How does he manage to be human at all, seeing
that the only link to the parents is a couple of germ cells each
so small that you need a microscope to see it ? To the extent
that this is a chemical question the solving of it lies entirely
in the future ; but the science of genetics has already gone a
long way by its own methods towards finding what sort of
controlling device is present in these germ cells, and biochemists are ready to follow up any hint of a chemical kind
that the geneticists may hit upon.

Dr. Eggleton has told us what the living cell contains, but *What is
its structure, and how does it " work " ?* Professor John E. Harris of

Bristol provides an answer to this question in the following extract. We shall see that the cell is bounded by a very thin membrane with remarkable powers of selecting what shall pass into and out of the cell ; that it contains a nucleus which not only controls the activities of the cell by its manufacture of enzymes, but also contains the chromosomes, those vitally important factors by means of which the individual character of the cell is handed on to its two daughter-cells when, by the usual process of " mitosis " the cell multiplies by dividing into two. Also that the remainder of the protoplasm in the cell, the fluid " cytoplasm " as it is called, has its own important functions and also contains certain structures such as the snakelike threads called " mitochondria " which seem to be essential for the life of the whole. The yolk of a new-laid egg is a single giant cell. Most cells are very much smaller ; the great majority indeed cannot be seen without the aid of a microscope, and our bodies contain many millions of cells, each with a life of its own, and each capable of living by itself if carefully transplanted into a suitable culture medium.

JOHN E. HARRIS

STRUCTURE AND FUNCTION IN THE LIVING CELL

THE protoplasm in which is carried on most of the important transformations of matter and energy associated with life, is in all higher animals divided into cells, each bounded by a cell membrane, a specialized region of the protoplasm which has an important part to play in controlling the nature and rate of transfer of substances passing into and out from the cell. Within the cell there is at least one very characteristic and permanent structure, the nucleus, bounded by a second membrane, the nuclear membrane, having similar properties to that of the cell membrane itself. The nucleus contains the chromosomes which are the ultimate source of the inherited characteristics of the organism, including among those the characteristics which determine the existence of life in the organism, as well as those determining what form that life shall take.

The last twenty years have witnessed a very great advance in our knowledge of the details of the structure and function of the constituents of the living cell—an advance which has been brought about by the use of a wide range of new techniques for the examination of the cell and its contents.

In describing some of the results of researches on the structure and function of the cell, it will be convenient to treat the cytoplasm, the nucleus, and the cell membrane separately, though the subjects will necessarily overlap at many points.

The Cytoplasm

The fact that protoplasm possesses a definite structure, an organized architecture in three dimensions, is amply evidenced by the complexity of the development of an egg into an elaborately proportioned and organized embryo and adult. It is inconceivable that this complex organism could arise from a random " mush " of chemical molecules in solution. Experiments with the centrifuge and micromanipulator show in fact that organized development can be interfered with by breaking up this *ultrastructure*, but as we have seen, it often reforms itself after such a breakdown.

This structure must be determined by the protein constituents of the cell, since it is only in these elaborate molecules that the possibilities exist for such an organized yet mobile meshwork.

The structure is not apparently resident in the visible cytoplasmic inclusions, for they can be removed from an unfertilized egg by ultracentrifugal treatment; such an egg, if fertilized, produces a normal embryo. Apart from the possible effect of the fertilizing sperm the organization must be present in the clear *hyaloplasmic* portion of the cytoplasm.

By the use of the Altmann-Gersh low temperature dehydration process, and independently by extracting a ground-up cell mush at low temperatures, the cell proteins have been shown to comprise three distinct components : a highly soluble fraction of globular protein molecules, a less soluble filamentous protein constituent, and a third fraction similar in physical properties to the second but much more insoluble. Only after removal of this third fraction does all trace of cell structure disappear from the residue ; this portion therefore includes the structural proteins. It is possible that most of the second component is of nuclear origin since it contains nucleic acid in the filamentous molecular form.

The existence of this semi-permanent hyaloplasmic mesh-work is confirmed by viscosity determinations on the living cell, which show that, though injected salt solutions diffuse freely through the cytoplasm, which therefore contains a continuous aqueous phase, protoplasm possesses the elasticity of a solid as well as the viscosity of a liquid. This combination of solid and liquid properties is characteristic of mesh-like aggregations of long molecules in colloidal solution. The presence of such long molecules is also proved by the bi-refringence of protoplasm stretched into filaments. The long molecules then snap together over small areas, producing patches with an orderly " crystalline " array.

Near the cell surface this crystalline arrangement of protein molecules is often much more marked ; in the cell membrane itself the protein molecules are present as a definite series of concentric sheaths.

The degree of crystallinity of the cytoplasmic proteins is also very high in contractile regions. Cilia, flagella, and muscle fibres are all highly birefringent, the long axis of the protein molecules lying along the length of the fibre. The birefringence decreases during contraction, suggesting a loss of the crystalline orderliness. Just as a crowd of people all standing straight up and shoulder to shoulder would stand higher on the average than a similar group crumpled into random attitudes and sprawled in every direc-tion, so the two states of the muscle molecules may represent the elongated orderly and the contracted disorderly condition.

The fact that a thick surface layer of the cytoplasm is birefringent prior to and during cell division is of considerable interest, since it suggests that cell cleavage is produced by active contraction of the cytoplasm and not by a change in surface tension of the cell membrane, as has often been supposed.

The contractile properties of protoplasm are exhibited not only in specialized parts of the cell—cilia, flagella, and muscle fibres, but in all parts of the cytoplasm. Cell division and protoplasmic streaming as well as the contractility of the fibrous elements above are all similarly inhibited by very high " hydrostatic " pressure, by oxygen lack, presence and absence of metallic ions, and by certain drugs.

The metabolic activities of the cytoplasm—respiration, assimilation, secretion, excretion, etc.—are all enzymic processes which depend on the continuous production of the enzyme molecules (themselves proteins) which is a continuous feature of the living protoplasm. Many of these processes are associated with the presence of cell inclusions.

The cytoplasm contains, in addition to the clear hyaloplasm, a very large number of such inclusions, varying in form, quantity and nature from one type of cell to another. Many of them are not directly connected with the living activity of the cell—crystals of excretory products, food reserves such as starch, glycogen and fat droplets, skeletal stiffening rods, etc. Others represent the end products of active secretory processes —enzyme granules of the gut cells, " droplets " of silk protein from the silk gland cells, etc. Two types of inclusion are, however, so universal in their occurrence and, in the cells which have been investigated, are so constant in their general composition that they are believed to be essential to the life of the cell.

The smaller of these two inclusions—the *microsome*—is at or below the limits of microscopic visibility, varying in size from 0.05-0.3μ ($\mu = \frac{1}{1000}$ mm.) in diameter. The microsomes can be isolated by ultracentrifuging a cell *brei* (or mush), and when in a concentrated form appear as an amber-coloured jelly. Their principal constituents are phospholipins (complex fatty molecules containing phosphoric acid and organic acids and bases) and nucleo-proteins (compounds of simple proteins, with nucleic acid). The *mitochondria*, the larger inclusions, which may occur as granules or elongated flexible threads, are usually $1-2\mu$ in diameter and may be up to 20μ in length. Their composition is strikingly similar to that of the microsomes, including phospholipin and nucleoprotein components as well as many other constituents. They appear to be self-perpetuating elements in the cytoplasm ; the thread-like types, for example, grow in length but not in diameter, and fragment transversely to produce new mitochondria. If most of them are removed from a cell, death usually follows.

In addition to the phospholipin and nucleoprotein constituents, it has been claimed that a very large number of important enzyme systems (including those involved in protein synthesis), with vitamin C and a number of other substances like cytochrome, concerned with oxidation processes, are present in the cytoplasm largely or entirely confined to the granules. Mitochondria occur in enormous numbers in cells which are actively manufacturing protein, such as enzyme secreting cells, silk gland cells, oocytes (developing egg-cells), and nerve cells, while they are fewer in cells which have a lower protein production such as those of heart, lung, and kidney tissue. It has therefore been suggested by Brachet in Belgium and by many other workers that the main function of these mitochondria is to provide a surface on which all the molecules concerned in protein manufacture can be highly concentrated under the optimum conditions for producing the protein. The protein-building enzymes, the requisite energy sources, and the amino-acid units for synthesis of these giant molecules, must come into very intimate contact. The constancy of the molecular pattern of the protein suggests an intricate mould or template perhaps provided by the nucleoprotein, against which this pattern can be built up. All these conditions are more likely to be achieved at an active surface than in free solution, and this may provide the *raison d'être* for the microsomes and mitochondria.

There is, incidentally, a considerable similarity of behaviour between the mitochondrion and a virus particle. The latter is endowed with the same powers of protein synthesis—in this case of self-reproduction—and the tobacco mosaic virus is largely if not entirely a nucleoprotein of the same type as that found in cell cytoplasm. When injected into the cytoplasm of a tobacco plant cell, the virus particle will go on reproducing itself indefinitely at the expense of the amino-acids and energy sources of its host.

A more elusive cytoplasmic inclusion is the so-called Golgi apparatus, a complex structure of water vacuoles partially sheathed with fatty material and revealed in a protean diversity of form by various complicated histological methods. Baker of Oxford has been largely responsible for determining its nature; it normally accompanies and plays some as yet undisclosed part in cell secretion, perhaps in the mobilization of cell reserves. It is often associated with the production of elongated fibrous protein molecules, as for example in the nerve cell and the spermatocyte, and its occurrence along with the centrosome from which the spindle fibres radiate also seems to fit in with this association. How far its function is necessary to the life of the cell is not decided—a comparable structure in Protozoa can be regenerated if it is removed by centrifuging.

THE CELL MEMBRANE

The cell surface is often protected by a non-living wall, of cellulose in plant cells and of various skeletal carbohydrates and proteins in animal cells. These reinforcements usually play only a minor part in the life of the cell, giving it mechanical protection, and on occasion enabling very high pressures to be developed by osmotic forces without the rupture of the cell. The real barrier to the passage of substances in and out of the cell is provided by a much more delicate membrane, often called the plasma membrane, since it can be looked upon as a surface modification of the cytoplasm itself. If not too extensively damaged, a punctured plasma membrane can be re-formed from cytoplasmic constituents—a process which apparently requires the presence of calcium ions in the fluid surrounding the cell.

It has long been known, from work by Overton in this country and by Collander and Barlund in Finland, that the plasma membrane is highly permeable to substances which are soluble in fats and in fat solvents. This fact, coupled with the chemical identification of complex fats in the membrane of broken red blood corpuscles, suggested that fats were a prominent constituent of the cell membrane. The amount of fatty material in the corpuscle membrane is, however, surprisingly small—only enough to make a layer two molecules in thickness. This result was strikingly confirmed in America by a measurement of the electrical resistance of the cell wall to alternating current. Frickle found that the main resistance to the passage of ions through the cell was limited to an insulating layer at the cell surface which was about 50 Å in thickness ; if the insulating layer is provided by fat molecules this result agrees closely with the chemical estimate of the fat content.

The work of Danielli, Harvey, and others has shown that the stability of this layer is maintained in two ways. In the first place, though its fat molecules are free to move to a limited extent, they are confined to a pair of monomolecular layers, with the molecules oriented in a definite direction. Secondly, outside and inside of this fatty layer are protein molecules, probably linked together to form an elastic meshwork. The presence of these protein molecules confers a certain rigidity on the membrane and also reduces the high surface tension which would arise if the membrane were composed only of oriented fat molecules.

The plasma membrane is the main barrier to the free migration of materials into and out from the cell, and it is an extremely effective one. It has been calculated that it would take 10^{32} *years* for a single large protein molecule to pass through it, so that the escape of the essential cell proteins is totally prevented. For other types of molecule its permeability properties are equally well suited to the requirements of the cell. It passes small uncharged molecules like oxygen and carbon dioxide relatively rapidly, but charged ions only very slowly. In this way cell respiration can be adequately assured whilst the retention of salts is provided for. Water molecules pass through fairly rapidly, so that the cell often behaves like an osmometer. Large molecules, particularly those which are very soluble in water, penetrate very slowly— a surprising example is provided by sugars which are passed through cell walls by an elaborate secretory process which accelerates the passage considerably. Fatty substances penetrate cells rather more readily—it is significant that many anaesthetics are fat solvents and act on cells very rapidly. The problem of the penetration of large molecules of fat (vitamin A) or protein (the hormone insulin) is not easy to solve, for both these substances have important effects on the cell ; it has been suggested that the cell membrane is a complex mosaic, small areas of it possessing different permeability characteristics. There is experimental evidence that this is true for the entry of glycerine molecules, and probably also for other substances.

The cell membrane has an additional function to perform besides that of a selective barrier to the passage of substances. It is the receiver and transmitter of stimuli between adjacent cells—a process which is highly developed in the specialized nerve cells. The origin and transmission of this stimulus certainly takes place in the surface of the cell and is associated with a definite change in the electrical potential across the two sides of the plasma membrane. Since externally applied potentials of the same order will themselves " discharge " the cell membrane, the propagation of a nerve impulse is a self-exciting process, each section discharged stimulating the

adjacent area, and so on. If the mechanism can reset itself
after the discharge has passed, the condition is typical of that
of the nerve cell, but Blinks in the U.S.A. has described plant
cells where the " resetting " does not take place, and the
membrane remains in the discharged condition. It may be
that this process is the basis of some special types of cell
response, as, for example, in the activation of an ovum on
fertilization, but this has not been directly proved.

Mention has been made of the possible mosaic structure
of the cell membrane. The transfer of substances across
the cell, as in the gut or in the kidney tubule, is often achieved
by mechanisms of much greater complexity. The cytoplasmic
metabolism plays a very definite part in the intake, transfer,
and output of substances, often by " secretory " mechanisms
of which we are largely ignorant. Glucose, for example, can
be collected from the urine in the kidney tubule and passed
back into the blood even though it is there at a higher concentration than in the urine.

The Nucleus

Experimental evidence of the function of the nucleus
dates from the end of the nineteenth century, when it was
shown by Lillie to be essential for the regeneration of a fragment of a protozoan (*Stentor*); Townsend also found that
without a nucleus a plant cell fragment was unable to form a
cellulose wall. The greatest development of our knowledge of
nuclear function, however, undoubtedly dates from the discovery of the chromosomes and the part played by them in
heredity. This is so familiar a story that it is unnecessary to
repeat it here—but it should be realized that for many years
after the discovery of genes we were still completely in the
dark as to how these unit characters acted to produce their
visible effects.

Scott-Moncrieff, working on flower pigments, showed
that genetical differences in flower colours were associated
with small clear-cut chemical differences in the anthocyanin

pigments, differences which could clearly be traced to simple enzyme mechanisms. The connexion between some genes and cell enzyme systems was therefore postulated. Many years later, in the U.S.A., Horowitz, Beadle, and Tatum showed that in cells of the mould *Neurospora*, which can normally be grown in a completely inorganic culture medium, certain heritable varieties existed which were incapable of carrying out specific chemical steps in the manufacture of the amino-acids, the first stage in the building up of the cell proteins. It is now clear that *all* of the vital activities of the cell are controlled by enzyme mechanisms, and that the ultimate source of these enzymes is to be found in the gene molecules present in the chromosomes. The fact that a mutation does not occur as a gradual change in the gene but as a sudden jump, and that it can be artificially produced by a single quantum of energy suggests that the gene behaves as a single molecule ; its size limits appear to be within the range of known protein molecules.

Cytological work on fixed and stained cells has provided us with a very clear and detailed picture of the process of chromosome division—though we are still largely in the dark as to the origin of the forces involved in mitosis. Equally important is the knowledge of how the gene produces and liberates its enzymes into the cell cytoplasm, and on this point some useful suggestions have been put forward.

The nucleus of all cells contains a considerable amount of nucleic acid, a complex molecule containing a pentose sugar, phosphoric acid, and a series of organic bases. This acid unites with basic proteins to form nucleoproteins which, as a whole, are still acid in nature and in the nucleus can therefore be coloured by basic dyestuffs—this property has given the material the generic name of *chromatin*.

There are two different types of nucleic acid, differing in the nature of the sugar molecule and in one of the organic bases forming part of the structure. One of these—thymo-nucleic acid—is almost confined to the nucleus, and gives rise to chromatin, the other—ribonucleic acid—is found in the cytoplasm, where it forms nucleoproteins which do not stain with basic dyestuffs but can be detected by another cyto-chemical colour test with the dye pyronin.

The pyronin test shows that ribonucleic acid occurs not only in the cytoplasm but also in the nucleus, particularly in the *nucleolus*. This small nuclear inclusion occurs in most cells and is particularly prominent in cells manufacturing large quantities of protein (nerve cells, oocytes and many gland cells).

The amount of chromatin in the nucleus rises and falls during the cycle of nuclear division ; the chromosomes stain most heavily in the mitotic stage. Inverse changes occur in the ribonucleic acid content of the nucleolus, and an intimate connexion of the two nucleic acid cycles is suggested by the fact that in many cells there is a nucleolus attached to each chromosome.

Ribonucleic acid, in the cytoplasm and in virus nucleo-proteins, is known to be an essential component of the protein manufacturing system ; it therefore appears reasonable to suppose that the thymonucleic acid in the chromosome has a similar part to play. The particular form of protein manu-facture in which the chromosome is concerned is its self-repro-duction ; it may be that the difference between the two nucleic acids is connected with the limited protein production involved in the single division of the chromosome into two chromatids, one of which will be ultimately passed on to the daughter-cell.

We may summarize the facts described above by regarding the cell as an extremely complex chemical workshop ; the entry of raw materials through the membrane is in part controlled by the properties of the membrane itself, in part by the changes in the cytoplasm which determine the quantity and form of the demand for external supplies. In the cyto-plasm are contained the enzyme systems which are the chemical " machine tools " turning out appropriate cell products ;

some of these machines also provide for the chemical transformations which release the energy to " drive " the mechanism.   The supply of enzymes appears to be maintained by
the nucleus, which turns out master copies which are duplicated in the cytoplasm.   This somewhat crude simile serves
to emphasize still another feature of cell metabolism about
which we know little—the maintenance of an orderly succession
in the multitudinous processes taking place side by side in
the protoplasm.   Biochemical studies on the carbohydrate
metabolism of muscle and on other cells have thrown some
light on the manner in which the successive steps are linked
in a definite order, but an enormous field of cell physiology
and biochemistry still remains to be explored.

As this brief account has shown, biologists, physicists, and
chemists of all nations have played a part in building up the
outlines of the present picture ; on its further elaboration
and understanding depends much of our future progress.
The conquest of cancer and of other cell diseases ; a fuller
understanding and possibly the influencing of the processes
of inheritance ; and the success of all other human efforts
which require the control of living activities of organisms,
will measure the extent of the future advance of knowledge
in this field.

" Enzymes " have been mentioned several times in Professor
Harris's lecture, and it has been suggested that some sort of enzyme
action is the secret of life. *What are enzymes and why are they so
important?* An answer to this question is provided by Prof. D. W.
Ewer of Rhodes University. We shall find that enzymes are proteins
with very specialized powers of producing chemical changes in certain
other molecules without being changed themselves. They are the
" catalysts " of the living organism.

In the digestion of our food, for example, they play a vital part.
All the starches, fats, and proteins, and most of the sugars in our food
consist of molecules far too large to pass through the walls of the
digestive tract. So they are split up by the enzymes of the saliva and

the other digestive juices, each enzyme being a specialist in performing one single operation. As a piece of bread is being chewed, for example, an enzyme called ptyalin makes the first attack on the molecules of starch. This attack continues in the stomach, where conditions are still acid, but after leaving the stomach the acidity is neutralized, and then other enzymes take up the attack on the molecules (now sugars instead of starches) left by the ptyalin. In each process the enzyme joins a molecule of water to the molecule of starch or sugar, and then splits the whole neatly in two, almost like a man splitting logs with a wedge. The enzyme is like the wedge ; it makes the operation easy but is not affected itself.

The enzyme, however, is much superior to the wedge in one respect. It can usually, under suitable conditions, work as well in the opposite direction and join the split logs together again. The actions produced by enzymes are reversible, and that is one reason why they are able to play such an important part in the working of all living machinery. These ideas are developed more fully in Dr. Ewer's essay, and in it, too, he introduces us to the action of hormones and of genes. Both of these important matters are dealt with again later on in this volume, and any temporary gaps which may exist in the picture for some of our readers will then be filled up.

DENIS WILLIAM EWER

THE BIOLOGY OF ENZYMES

TO-DAY everyone is so fully aware of the dangers of bacterial infection that if he cuts his finger he will in all probability treat the cut with disinfectant. One disinfectant which might be used is hydrogen peroxide. In the bottle this is a harmless looking liquid, but when placed on a cut surface it froths violently. What has happened ? Hydrogen peroxide is a simple compound that can be broken down to water and gaseous oxygen. Normally this decomposition goes ahead very slowly, but when the compound is placed on a cut tissue the reaction takes place very rapidly. Oxygen gas is set free and causes the characteristic frothing and, incidentally, the disinfection. Some substance in the cut tissue has greatly speeded up the decomposition of the hydrogen peroxide, and this substance we call an *enzyme*.

Potatoes are composed almost entirely of starch and water. When potatoes are eaten the starch is broken down in the alimentary tract to sugar. Outside the body this decomposition hardly takes place at all in water at body temperature. If, however, we add some of the digestive juices from the intestine to the water the starch is fairly rapidly decomposed. Clearly the digestive juices contain some substances which speed up this decomposition, and these substances are also enzymes.

This acceleration of chemical reactions which normally proceed slowly is a well-known phenomenon in chemistry and has great industrial importance. The chemists call such substances *catalysts*. Enzymes are catalysts elaborated by living organisms. There is an enormous variety of enzymes in an animal's body and each is capable of accelerating only one type of chemical reaction. The enzyme in the tissues which breaks down hydrogen peroxide will have no effect on starch, nor will the starch-splitting enzymes decompose hydrogen peroxide. This specialization of enzymes, or their *specificity*, as it is usually called, is used as a method of classifying them. Thus all types of enzyme which break down protein molecules are called proteinases, all enzymes which break down cellulose are called cellulases and so on.

What sort of things are these enzymes? As has been stated above, the enzymes of the digestive juices are capable of exerting their effects away from the animal's body, and in fact the majority of enzymes may be successfully separated from animal tissues and still remain active. There is therefore nothing mysterious or " vital " about their activity and it has been possible to purify and crystallize many different sorts. This purification has allowed the chemists to show that enzymes are protein compounds of high molecular weight. For example, the molecular weight of water is 18 and that of common salt 60, but the molecular weight of pepsin, the main enzyme found in gastric juice, is 35,000, while urease, an enzyme extracted from plant tissues, has a molecular weight of 480,000. Frequently it is possible to separate from the protein molecule a complex, characteristic non-protein fragment which is spoken of as a " prosthetic group ". This prosthetic group in some enzymes contains a very small percentage of metal such as iron, copper or zinc.

Since it is possible to isolate many enzymes in a pure state it is possible to obtain a measure of their efficiency. This has been done in a number of cases, and the results indicate that they are highly variable in performance. One of the most sluggish so far recognized occurs in the liver and there oxidizes certain types of amino-acid. One molecule of this enzyme will oxidize about 35 molecules of amino-acid in one second at blood heat. The most efficient yet found is the one which decomposes hydrogen peroxide. At freezing point one molecule of this enzyme will in one second decompose 44,000 molecules of peroxide. But these extremes do not reflect the average performance, which seems to be about 300 molecules per molecule enzyme per second.

How do enzymes bring about this great speeding up of chemical reactions ? There is little doubt that the enzymes combine with the various compounds which are involved in the reactions they accelerate. In a few cases a direct demonstration of this combination is possible. There is a big class of enzymes called oxidases. These are able to bring about the oxidation of fairly simple chemical compounds by molecular oxygen. These enzymes can exist in two distinct conditions, one oxidized and the other reduced. In certain cases the two conditions may be recognized by their characteristic spectra which are due to their prosthetic groups. If such an enzyme is mixed with the substance, or, as it is usually called, *substrate*, which it will oxidize, and a liberal supply of oxygen is made available, then the prosthetic group shows the spectrum of the oxidized form. If, however, the oxygen supply is cut off, the reaction will stop when all the oxygen is used up. The spectrum of the prosthetic group then corresponds to its reduced condition. If oxygen is readmitted, the oxidized type of spectrum reappears.

It is clear from this kind of experiment that the substrate takes oxygen from the prosthetic group, so that the latter is reduced and the substrate oxidized. In the presence of oxygen, as soon as the substrate reduces the prosthetic group, the latter is reoxidized by the molecular oxygen. This process of alternating oxidation and reduction will go on until all the substrate is oxidized and none is left to reduce the prosthetic group. It will be obvious from this that a very little enzyme is capable of oxidizing an unlimited quantity of substrate and that the more enzyme present the faster will the substrate be oxidized. Both these conclusions have frequently been confirmed by experiment.

In the case we have considered, it seems easy to understand the way in which the enzyme is involved in the reaction which it catalyses. In fact the events must be more complicated than this, for, although we can follow the oxidation and reduction of the prosthetic group easily, the whole enzyme molecule is concerned in the process. The prosthetic group alone will not act as an enzyme. In other cases, the nature of the chemical reactions concerned is not easily demonstrated, and this aspect of enzyme chemistry is a subject of intensive research and controversy at the present time.

In almost all the reactions which we have so far described, the enzymes have been accelerating reactions in which the substrates have been broken into smaller pieces. Enzymes are capable of accelerating synthetic as well as degradative reactions. All chemical reactions are reversible, but in many cases this reversibility is so slight as to be negligible. In other cases the reversibility is more clearly marked. For example, if a certain fat is mixed with a fat-splitting enzyme, a lipase, the fat will be split up until only about 40 per cent. of it remains. Similarly, if the breakdown products are mixed together with the enzyme, fat will be formed ; but the quantity of fat formed will never exceed about 40 per cent. of the total possible. The enzyme can speed up both reactions.

One surprising result of this reversibility has fairly recently been discovered. It has been realized for a long time that plants are capable of abstracting carbon dioxide from the air and turning it into sugars and starch. The green pigment chlorophyll plays an essential part in this process. It has now been shown that in animals carbon dioxide can also be used by the liver in building up sugars from simpler materials. This has been demonstrated by injecting into mice a salt of carbonic acid in which the carbon atoms were radioactive. After an interval the mice were killed and from their livers was extracted glycogen, a compound very similar to the starch of plants. This glycogen was found to be radioactive. Some of the carbon atoms in the carbonic acid had become incorporated in the glycogen. To get there they had first to be built into sugar molecules. So it appears that " carbon assimilation " can also occur in animals. The chemistry underlying this event is complex, but it seems probable that it is simply the fortuitous result of the reversibility of enzymic reactions. Reactions which normally produce carbon dioxide will, because of their reversibility, also incorporate carbon dioxide. Some of the radioactive carbon is incorporated in fairly simple molecules in this manner and then, owing to the reversibility of a further long chain of reactions, a small number of the radioactive atoms find their way into sugar molecules. Whether this particular phenomenon has any biological significance we have yet to discover.

The capacity of enzymes to accelerate synthetic reactions is of immense importance in the building up of our bodily structures from the products of digestion. The determination of the conditions which make for synthesis or for degradation is of the greatest importance to our understanding of many biological phenomena. Recently a French scientist, Roche, has claimed to have isolated an organic substance which augments the synthetic activity of an enzyme, but not its degradative action. If this type of compound is common it may considerably assist us in understanding the way in which cells control their complex chemical processes.

These, then, are some of the more important properties of enzymes. Their activities are manifold. They are concerned in the normal processes of digestion; in the oxidation of food material within the cells of the body; in the building up of reserve stores in the liver, muscles and fat depots; in the elaboration of various special chemical compounds such as hormones, in short in all the major chemical activities of the body. They are also concerned in events less obviously chemical. The light of glow-worms depends upon an enzyme-accelerated chemical reaction; the venoms of vipers and cobras depend for their actions on enzymes; there is good reason to believe that the contraction of muscle is closely dependent on enzyme action, while certain scientists believe that enzymes play a very important rôle in the conduction of nervous impulses. It is therefore not surprising that many important aspects of biology find a common meeting-ground in enzyme action. Three of these—nutrition, endocrinology and genetics—are discussed below.

Enzymes and Nutrition

As was pointed out above, it has proved possible to separate from many enzymes characteristic chemical groupings which are referred to as prosthetic groups. These groups frequently contain one or a few metallic atoms such as those of iron, copper, zinc, or manganese. While the importance of iron in the diet has long been realized, the importance of very small quantities of other metallic elements has only fairly recently been recognized. Normally there is no deficiency of these elements, but under exceptional conditions deficiency diseases may arise. The importance of these " trace elements " in plants has been described in an article by Professor Stiles.

In the laboratory animals may be reared on diets deficient in some particular metallic element and the effects of this deficiency studied. For example, it has been shown that a diet deficient in manganese will retard the growth rate of young rats and produce irregularities in the functioning of the reproductive organs. Manganese occurs in the prosthetic group of an enzyme which is concerned with the elaboration of the excretory product, urea. In the absence of this metal in the diet the young animals will not be able to synthesize the enzyme. Clearly the absence of such an important enzyme is likely to cause trouble, but how far the various aspects of the manganese deficiency syndrome in rats are to be attributed to the lack of this enzyme awaits investigation. There is some evidence that the disease in hens known as perosis may be due to manganese deficiency. Similar experiments have shown that rats fed on a zinc-deficient diet grow poorly. The rats in these experiments had to be kept in special cages made of monel metal instead of the usual ones made of galvanized iron in case the rats should make good the deficiency of zinc by gnawing the bars of their cages. Zinc occurs in an enzyme which ensures the rapid liberation of carbon dioxide from the blood during its passage through the lungs. However, lack of this enzyme may not be the cause of retardation of growth, as zinc also occurs in another very important compound, namely insulin. Thus the poor growth rate of rats on a zinc-deficient diet may be due to disturbed sugar metabolism rather than anything else.

Deficiency diseases may then arise if diets do not contain the necessary metallic atoms needed to build up the prosthetic groups of their enzymes. In the prosthetic groups of some enzymes there are also characteristic configurations of atoms which most mammals appear to be unable to put together for themselves. They find it necessary to obtain these in their food. They are members of that class of nutritional essentials which we call vitamins. Thus the two vitamins riboflavine and niacin are constituents of the prosthetic groups of enzymes concerned with the energy-liberating oxidations which take place in all tissues. A third example is thiamin, the vitamin which prevents beri-beri. This vitamin is part of the prosthetic group of an enzyme responsible for oxidations in various tissues, including the brain.

Some of the vitamins, then, are chemical substances required by the mammalian body to build up the prosthetic groups of some of its enzymes. The mammal is incapable of synthesizing these things. During recent years a number of studies have been made on the vitamin requirements of lower animals and particularly of the insects and protozoa. The interesting fact has emerged that many of those substances which are necessary supplements to the diet of a mammal must also be added to the diets of insects and the higher protozoa if these are to grow properly. In fact quite recently a new B vitamin, apparently peculiar to the meal worm beetle, has been recognized and named vitamin B_T, the T standing for *Tenebrio*, the Latin name of the beetle. Clearly, like the mammals, these animals also lack the capacity to synthesize these substances, and we may presume that they possess enzymes with similar prosthetic groups and similar functions to those of the Mammalia.

However, not all protozoa require these vitamins, for some possess the means, that is the enzymes, to synthesize them. An especially interesting case has been worked out by the French biologist Lwoff. He has shown that in one species of protozoan, *Polytoma*, the vitamin thiamin can be synthesized completely from inorganic material The synthetic process, which is complex, has the following pattern :

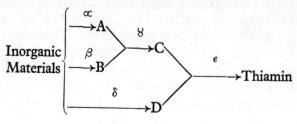

A, B, C, and D are various organic intermediates in the synthesis, while α, β, γ, δ, and ϵ represent enzymes responsible for the various syntheses. As long as the animal has all five enzymes it has no need for thiamin as a vitamin, for it can make it itself. However, not all species of *Polytoma* have the full complement of enzymes. One species, for example, possesses only the enzymes δ and ϵ. This protozoan must have C in its diet to make thiamin. C is a vitamin for this species ; thiamin is not, although of course thiamin can replace C. Lwoff has investigated many other cases of loss of synthetic enzymes and emphasizes that this loss of synthetic ability is characteristic of the physiological evolution of animals. The plants, as a whole, have not lost synthetic ability. It is tempting to suggest, but the idea must be regarded as purely speculative, that the greater diversity of function characteristic of animals may have been made possible by this loss of synthetic ability. We shall shortly see that there is a close relationship between genes and enzymes. One can imagine that when animals started along the path of dependence for their vitamins and other food materials on the synthetic ability of plants they were free to modify those genes which had previously been essential for these syntheses. These modified genes were able to serve completely new purposes. The gradual specialization of these genes may have been the foundation of the great differences which exist to-day between the higher plants and animals.

ENZYMES AND HORMONES

The second large class of biological compounds which has long attracted general interest is the hormones. Until recently the ways in which hormones produced their effects was completely obscure. During the last few years, however, very interesting results have been obtained by Drs. C. F. and G. T. Cori and their collaborators in the United States. For many years these workers have studied the changes undergone by carbohydrates in the body, and in 1947 they were both awarded Nobel prizes, together with Professor Houssay (of whom a little later), for their many notable discoveries in this field. One of the main events in the metabolism of sugar in the body is its storage in the liver as glycogen. There are a number of steps in the formation of glycogen from the sugar, glucose, and each step has a presiding enzyme. They are as follows:

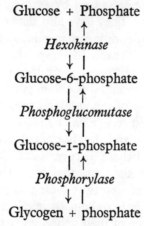

All these reactions are reversible.

It will be remembered that in pancreatic diabetes there is an excess of sugar in the blood. Some of this sugar comes from the breakdown of the liver glycogen. If the hormone of the pancreas, insulin, is injected into a diabetic the blood sugar level is temporarily restored to normal. Professor Houssay has made the interesting observation that if you remove both the pancreas and the anterior lobe of the pituitary gland from an experimental animal no diabetes results. It would appear that a hormone (or possibly hormones) from the pituitary has the opposite effect to that of insulin, namely that it increases the level of sugar in the blood; so that if the sources of both insulin and the pituitary hormone are removed,

the blood sugar level remains about normal. Confirmation
of this is obtained by removing the anterior pituitary gland
alone. The blood sugar level then drops as the insulin from
the pancreas is given free rein.

The workers in the Coris' laboratories have made a number
of interesting observations on the enzymic aspects of these
effects. They have succeeded in preparing purified extracts
of the enzyme hexokinase, and have been able to show that
its action is strongly depressed by extracts from the anterior
lobe of the pituitary. They have also shown that this pituitary
inhibition is offset by addition of insulin, and in a further series
of experiments that insulin increases the capacity of phos-
phorylase to convert glucose-1-phosphate to glycogen. In the
action of these hormones on these particular enzymes we
seem to see the basis of the antagonism of the two endocrine
glands which was first recognized by Houssay in 1931. Carbohydrate metabolism and its hormonal control is a highly
complex subject and these discoveries are far from solving all
the problems which exist ; but they are of the greatest moment
as they give us for the first time some real insight into the way
in which hormones bring about their effects and, in this case
at least, it is by virtue of their action on enzymes.

ENZYMES AND GENES

It has already been hinted that there is a relationship
between genes and enzymes. Occasional cases which illustrated this have been recognized in the past. For example,
albinism in man is known to be an inherited character. It
may be shown to be due to the absence in the skin of an
enzyme which forms melanin, the pigment of skin and hair.
In the normal individual, with the normal genes, this enzyme
is present, and we may assume that there is some link between
the presence of the normal gene and the formation of the
enzyme. Another similar example is a hereditary disease
known as alcaptonuria. In this condition the urine turns

black on standing. This blackening is due to the oxidation of a substance called homogentisic acid. Normally this is completely oxidized in the body to carbon dioxide and water, but when the normal genes are absent and those associated with alcaptonuria are there instead, the enzyme initiating this oxidation is lacking. Again it appears that the normal gene is concerned in some way with the production of an enzyme.

Our knowledge of the relations between genes and enzymes has been greatly increased during recent years. The experiments have been chiefly made on moulds, yeasts and protozoa. Outstanding contributions have been made by the American geneticist Beadle, who has studied the bread mould *Neurospora*. Mutations of the genes in the spores of the mould have been induced by X-rays or ultra-violet light. The mould can normally be cultured in a medium which is chemically very simple. The induced mutations which have been studied are those in which the culture medium has to be enriched by some normally unnecessary chemical before the mould will grow. That is to say, the mould has lost some synthetic capacit owing to the mutation. This loss of synthetic capacity has as its immediate cause the lack of an enzyme which occurs in the normal mould. A considerable number of such hereditary nutritional defects have been found and have fully established the close link which may exist between genes and enzymes. These experiments have also told us a great deal about the chemical processes which go on in *Neurospora*. They prove to be remarkably similar to those in our own bodies.

It is, however, necessary to sound a note of warning. Not all genes should be assumed to produce their effects by their ability or inability to initiate the synthesis of a particular enzyme. Many mutations in *Neurospora* affect its form of growth rather than its nutrition. These have, so far, not been studied. From the study of similar types of mutation in the fruit-fly *Drosophila*, and the Gypsy Moth *Lymantria*, there is evidence to show that certain genes produce their effects by modifying the rates at which various processes occur during

development and growth. There is the possibility that these
genes are concerned with the production of substances which
alter the activities of enzymes in the same sort of way as the
Coris have found various hormones alter the activities of the
enzymes of glycogen synthesis. This, however, is only a possibility and as yet we have no evidence about the way in which
these genes produce their effects. They may not be associated
with enzymes at all.

Reverting once more to those genes which are associated
with the production of particular enzymes, recent work on
bacteria and yeasts has shown that the picture is not as simple
as it might at first appear. Certain enzymes always depend
for their formation on the presence of the corresponding gene.
Their production is both initiated and maintained by the
relevant gene. In other cases an enzyme may be formed under
the influence of a gene, but once it is formed it will be produced
by the cell regardless of whether the gene is present or not.
If the gene is absent and all the enzyme is lost, it cannot be
regained. The production of this type of enzyme, then, is
gene-initiated but not gene-maintained. These findings,
however, belong to a chapter which is still being written, but
which when complete will certainly bring us nearer understanding how enzyme-initiating genes produce their effects,
and may also throw light on the difficult problem of how
enzymes themselves are synthesized.

Conclusion

In this brief survey an attempt has been made to show the
way in which a number of typical biological phenomena have
a common meeting-ground in enzyme chemistry. Enzymes
are clearly connected with so many vital functions that there
might be a temptation to say, " These are the secrets of Life ".
But they are not. Isolated and crystallized, they are as dead
as common salt. Together with the other constituents of cells,
the whole displays those characteristics we associate with

living matter. The more we learn of enzymes, of their substrates and of their interrelations, the more we understand
about the way living matter is put together, and very occasionally we see glimpses of how it " works ". But we will never
extract the " Secret of Life " from a cell any more than we
could extract the " Secret of Motion " from a steam engine.
The motion of the engine no more resides in the steel of the
piston rod than the life of the cell lies in its enzymes ; but the
life of the cell does lie in the spatial and dynamic relationship
of its parts, and, among these, enzymes are unquestionably of
the greatest importance.

Dr. Ewer has suggested that the differences between plant and
animal forms of life developed as certain organisms lost their store of
food-building enzymes, and developed other abilities instead. Among
these other abilities was a much-improved way of transmitting a
stimulus from one cell to the next—a very important matter in a many-
celled creature. Just as speedy and efficient communications are
essential in a modern state so are they a prime necessity in a many-
celled animal or plant.

The use of this word " cell " to describe the " bricks " out of which
living things are made is rather unfortunate in that it suggests a solid-
walled structure. It was a natural word for Hooke to use when, in 1667,
he peered through one of the early microscopes at a piece of dead cork,
and noticed its cellular structure. It is true that plant cells are usually
divided by walls of cellulose, but even these walls can be extremely
fine in the living plant, while the walls of animal cells are usually of
incredible thinness. Animal cells thus made the transmission of stimuli
easy and rapid even before they invented the special nerve-cell. But
some plants achieved an equal speed. Sir Jagadis Bose was the
pioneer in the study of the transmission of stimuli in plants, and two
brief extracts from his lectures are now given. By means of instruments
measuring ultra-microscopic movements he was able to show that
plants responded to changes in environment, and " even to the slight
fluctuations of light caused by a drifting cloud ". *How was a system
of internal communications demonstrated in plants ?*

SIR JAGADIS CHANDRA BOSE

THE WONDERS OF PLANT LIFE

IN plant life a stimulus takes a certain time before it gets a response. This stimulus may be of different forms, *e.g.* it may be a sound stimulus, a light stimulus, an electric stimulus, and so on. The feebler the stimulus, the greater is the time it takes to elicit the response. For instance, if one is called by a distant voice, one doubts whether he has been called at all, but in the case of a piercing scream, he starts up at once.

Now, the difficulty is that when the stimulus, the blow, is so strong as to get an instantaneous response, how is one to measure this infinitesimal time between the blow and the response ? And this must be done absolutely free from any personal interference, so as to ensure correct results.

After deep thought and careful experiments and researches of several years I invented and manufactured a highly sensitive instrument which could automatically record the " response time " of a plant even to one-thousandth part of a second. In order to convey a graphic idea of the principles under which it worked, I have made, by means of a few simple things, a crude form of the instrument, which will help us to form a clear idea of how a shock given to a plant would be recorded automatically by the apparatus by means of dots on its writing-pad, and also how to ascertain the exact time each plant took to respond to the stimulus received. Thus the plant now records its own history unerringly by its own hand as it were. And that the same results are obtained each time the experiment is repeated under similar conditions, shows that this recording of the response-time is a scientific phenomenon.

As an example of the similarities of reactions in plant and animal, I will describe the rhythmic activities of certain plants, in which automatic pulsations are maintained as in the animal heart. This phenomenon is exemplified by the telegraph plant, which grows wild in the Gangetic plain ; its Indian name is *bon charal* or " forest churl ", the popular belief being that it dances to the clapping of the hand. There is no foundation, however, for this belief. It is a papilionaceous plant with trifoliate leaves, of which the terminal leaflet is large, and the two lateral very small. Each of these is inserted on the petiole by means of a pulvinus. The lateral leaflets are seen to execute pulsating movements which are apparently uncaused, and are not unlike the rhythmic movement of the heart to which, as we shall see later, their resemblance is more than superficial.

In the intact plant, under favourable conditions, these movements are observed to take place more or less continuously ; but there are times when they come to a standstill. For this reason and because of the fact that a large plant cannot easily be manipulated as a whole and subjected to various changing conditions which the purpose of the investigation demands, it is desirable, if possible, to experiment with the detached petiole, carrying the pulsating leaflet. The required amputation, however, may be followed by arrest of the pulsating movements. But, as in the case of the isolated heart in a state of standstill, I found that the movement of the leaflet can be renewed, in the detached specimen, by the application of internal hydrostatic pressure. Under these conditions, the rhythmic pulsations are easily maintained uniform for several hours. This is a great advantage, inasmuch as in the undetached specimen the pulsations are not usually found to be so regular as they now become. So small a specimen, again, can easily be subjected to changing experimental conditions, such as the variation of internal hydrostatic pressure and temperature, application of different drugs, vapours, and gases.

Under varying conditions the same plant has been observed to take different response-times, as for instance, less in heat than in cold, less in summer than in winter, less in the morning than in the evening, and so forth. Again, different plants have different response-times. It is a remarkable fact that the mimosa is ten times as sensitive as a frog in giving the response.

And the idea that plants are of a lower order than animal life will cost many a sad disappointment.

As it has been the earnest endeavour of scientists in these experiments to minimize material friction in order to get the best results, so in our human concerns, it should be our best aim to minimize friction—which is, Ignorance.

Can a plant feel? Wordsworth said it was his belief that " every flower enjoys the air it breathes ", and Bose's experiments certainly seem to suggest a similar thought. But no trace has been found of a sense organ or of a nervous system in any plant, so that we must abandon the happy thought of countless acres of green fields thrilling to the sunshine of springtime—and incidentally the corresponding, painful thought of those acres dying of thirst in a drought or writhing with pain under reaper's knives or grazing animals. Plant movements are extremely slow because, rooted in the ground as they are, and drawing their food from the soil and from the air around them, they have no need for special, rapid movements. The ordinary movements of growth are sufficient. Thus if the tip of a root, or an opening bud, is photographed every fifteen minutes, and the subsequent film is run through a cine-projector at the usual rate of some fifteen pictures a second, the movements look just like those of an animal.

The stimulus which passes along animal nerves is at least partly electrical in nature, and Bose succeeded in demonstrating, by means of his sensitive apparatus, that rhythmical electric changes occurred in the stems of plants.

SIR JAGADIS CHANDRA BOSE

HOW PLANTS CAN RECORD THEIR OWN STORY

WE have now to consider the claims made by those who profess to discriminate character by handwriting. As to the authenticity of such claims, scepticism is permissible ;

but there is no doubt that one's handwriting might be modified profoundly by conditions, physical and mental. There still exist, at Hatfield House, documents which contain the signature of the historical Guy Fawkes. A photograph projected on the screen shows a sinister variation in those signatures. The crabbed and distorted characters of the last words which Guy Fawkes wrote on earth tell their own tale of that fateful night. Such is the tale that might be unfolded by the lines and curves of a human autograph. Can plants be made similarly to write their own autographs revealing their hidden story ? Storm and sunshine, the warmth of summer and the frost of winter, drought and rain, come and go about the plants. What subtle impress do they leave behind ? How are the invisible internal changes to be made externally visible ?

I have succeeded in devising experimental methods and apparatus by which the plant is made to give an answering signal, which is then automatically recorded into an intelligible script. The results of the new investigations are so novel that I spent several years in perfecting automatic instruments which completely eliminate all personal equations. The plant attached to the recording apparatus is automatically excited by a stimulus absolutely constant, making its own responsive records, going through its period of recovery, and embarking on the same cycle again without assistance at any point from the observer. A most sensitive organ for perception of a stimulus is the human tongue. An average European can by his tongue detect an electrical current as feeble as six micro-amperes, a micro-ampere being a millionth part of a unit of electrical current. I found that my Hindu peoples could detect a much feebler current, namely, 1.5 micro-amperes. It is an open question whether such a high excitability of the tongue is to be claimed as a distinct advantage. But the fact might explain the eminence of my countrymen in forensic domains ! The plant, when tested, is found to be ten times more sensitive than a human being.

When the plant has a surfeit of drink, it becomes excessively lethargic and irresponsive. By extracting fluid from the gorged plant, its motor-activity is at once re-established. Under alcohol its responsive script becomes ludicrously unsteady. A scientific superstition exists regarding carbonic acid as being good for a plant. But my experiments show distinctly that the gas would suffocate the plant as readily as it does the animal. Only in the presence of sunlight can the effect be modified by secondary reaction.

By means of apparatus specially devised, pulsative plants are made to record their rhythmic throbbings. It is shown that the pulse-beats of the plants are affected by the action of various drugs and divers stimuli, in a manner similar to that of the animal heart. Perhaps the most weird experience is to watch the death-struggle of a plant under the action of poison. Turning from death to its antithesis, life and growth, we are shown how the latter can be made visible by means of the new appliances. The infinitesimal growth of a plant becomes highly magnified in these experiments.

When I commenced my investigations, original research in India was regarded as an impossibility. No proper laboratory existed, nor was there any scientific manufactory for the construction of a special apparatus. In spite of these difficulties it has been a matter of gratification to me that the various investigations already carried out at the Presidency College have done something for the advancement of knowledge. The delicate instruments seen in operation at this lecture, which have been regarded with admiration by many distinguished scientific men in the West, are all constructed at the College workshops by Indian mechanics.

It is also with pride that I refer to the co-operation of my pupils and assistants, through whose help the extensive works, requiring ceaseless labour by day and night, all have been accomplished. Doubt has been cast on the capacity of Indian students in the field of science. From my personal experience I bear testimony to their special fitness in this respect. An intellectual hunger has been created by the spread of education. An Indian student demands something absorbing to think about and to give scope for his latent energies. If this can be done, he will betake himself ardently to research into Nature, which can never end. There is room for such toilers who by incessant work will extend the bounds of human knowledge.

Before concluding I wish to remark on the fact that all the varied and complex responses of the animals have been fore-shadowed in the plant. The phenomena of life in the plant are thus not so remote as has been hitherto supposed. The plant world, like the animal, throbs with responsiveness to all the stimuli which fall upon it. Thus, community throughout the great ocean of life, in all its different forms, outweighs apparent dissimilarity. Diversity is swallowed up in unity.

We have seen the probable origin of life, and perhaps we have realized something of the complexity of the processes involved in keeping the flame alive even in the simplest organisms. This is only the beginning of the marvels. Next there has to be devised some mechanism for the handing on of the torch. If this mechanism is purely vegetative, like the spreading roots of couch-grass, multiplication can be swift, but there is no scope for change, no possibility of the development of higher forms. What Nature has invented, therefore, is a method which permits of occasional changes, or " mutations " as they are called, while in general it follows a " like father, like son " policy. *How are family resemblances handed down ?*

In the following lecture the late Professor Dendy, who was well known in Australia, New Zealand, and South Africa before he came to King's College, London, and who was for many years a very popular lecturer, sums up much that we have just read, and then introduces the two great allied themes of heredity and evolution—two aspects of the problem of life which are indissolubly connected. It is true that evolution implies the emergence or development of characters which have not been inherited, or at least not inherited to that degree or in that form, but evolution cannot work unless these new developments can be handed on to future generations. Dendy will give an outline of the mechanism of heredity, and will show how " the luck of the draw ", as we might call it, plays a large part in determining the character, and therefore the fate, of any individual.

ARTHUR DENDY

THE STREAM OF LIFE

ALL typical organisms—animal or vegetable—are composed of cells ; minute nucleated masses of protoplasm, existing either singly or in many-celled aggregates. These cells are capable of reproducing themselves by a process of division, and each of the higher organisms, with certain negligible exceptions, starts its life in the condition of a single cell which we call an egg or ovum, or, to use a more general term, a germ-cell.

Whatever may have happened in the far-distant past, at the present day, so far as we can see, every living thing is the product of some pre-existing living thing, the relation of parent and child holds good throughout the whole organic world, and when we come to analyse this relationship from the biological point of view we find that it is always essentially based upon cell-division. Leaving out of account, as we may legitimately do for our present purposes, the stages of protoplasmic evolution that precede the appearance of the nucleated cell, we may say that the cell is the unit of organic structure, that all organisms are built up of such units in somewhat the same way as a house is built up of bricks, except that the process of building in the living organism is one of cell-growth and cell-multiplication, while the bricks of a house are brought together and combined into a building by some external agency. This fundamental conception of organic growth leads to the still more fundamental conception of living matter as a continuous stream of protoplasm, starting with the first appearance of life on the earth and continuing to the present day with undiminished vigour ; but it is a stream which in the process of time constantly branches out in new directions, giving rise ever to more complex and more diversified types of plants and animals. It is the stream of life.

When once the protoplasm of the egg, or germ-plasm, as it is technically termed, has developed into the mature tissues and organs of the adult body, it cannot, usually at any rate, be turned back again to germ-plasm ; it continues to live for a time, but the stress and strain of life gradually exhaust its vitality ; for a time, tissues and organs may be renewed, but ultimately some essential part of the mechanism of the body is worn out beyond the possibility of repair, and the death of the entire organism inevitably follows.

What provision, then, is made for the next generation—who mixes the next batch of dough ? Here I am afraid our analogy breaks down, and it breaks down just because the germ-plasm, unlike the dough, is a living substance capable of increasing itself indefinitely by growth and multiplication. What happens is typically this—a part of the original germ-plasm of each generation is set aside, taking no share in the development of the body, but remaining in the condition of comparatively undifferentiated protoplasm, while continuing to increase and subdivide into germ-cells. It thus appears that the old idea that the hen produces the egg is scarcely correct—it seems that the egg produces the hen and at the same time more eggs, which are accidentally, as it were, included in the body of the hen. The constant succession of germ-cells, each produced by division of a parent cell, constitutes the only really continuous stream of living protoplasm. The bodies of individual plants and animals developing from the germ-cells may be compared to local and temporary overflows from the stream, which sooner or later dry up and disappear, or, in other words, die. This is Weismann's well-known doctrine of the " continuity of the germ-plasm ", and for our present purposes we may take it as substantially correct, in principle if not in detail.

The simpler living organisms, which, like the amœba, consist each of only a single cell, are exempt from death, because in them the stream of protoplasm forms no overflows ; it consists entirely of germ-plasm, and no differentiated bodies are formed, so that there is nothing to die, nothing which cannot go on reproducing itself indefinitely. Death is the penalty paid for a higher life, based upon a greater complexity of bodily mechanism.

In all that has been said hitherto, which must be already very familiar to most of you, we have been endeavouring to pave the way for the consideration of what is perhaps the most difficult and certainly the most vigorously discussed problem of biology—the problem of heredity. With regard to the single-celled organisms such as the amœba this problem scarcely exists. Division of the parent cell entails division of all that that cell possesses. The daughter cells resemble the mother simply because they are that mother divided into two equal and similar parts.

With the higher organisms, each composed, perhaps, of many millions of cells, differentiated into many different kinds, and building up the most diverse tissues and organs, the situation is very different. In such a case how can a single, apparently undifferentiated germ-cell, which has never taken part in the formation or in the activities of the body as a whole, and exhibits none of the features which characterize the tissue-cells—how can such a simple cell give rise by growth and multiplication to all the different kinds of cells, arranged in all the different tissues and organs, more or less exactly as in the parent ?

The development of such an infinitely complex organism as, for example, the human body, from a microscopic egg-cell of apparently simple structure, seems, indeed, a kind of miracle, and the more closely we compare parent and child the more miraculous does the result appear, for not only is there a general resemblance in all essential features, but there is very frequently also a particular resemblance in minute peculiarities, such as the colour of the hair or eyes, the contour of the features, and so on.

It would be claiming far too much to say that we have as yet arrived at any explanation of heredity—this marvellously accurate reproduction in the child of the most minute details of bodily and mental organization exhibited by the parent. But the explanation is, perhaps, after all, not quite so difficult as it seems at first sight. Let us go back to our loaves of bread and ask ourselves why one loaf resembles another. Why does the loaf that was baked on Tuesday resemble that which was baked in the same oven on Monday ? The answer is obvious. One loaf resembles another because it is made from the same kind of dough and subjected to the same kind of treatment. If you take a different kind of dough, or subject the same dough to a different treatment, you will get a different result—and, as every housewife knows, there may be a vast difference between the loaves turned out by different bakers. The characters of the loaf clearly depend upon two sets of conditions : first, the nature of the dough itself, whether, for example, it is mixed with yeast or baking powder, water or milk, salt or sugar, and so on ; and, secondly, the nature of the treatment to which the dough is subjected, the shape of the tins in which it is baked, the temperature of the oven, and so forth. If all the conditions are accurately repeated for successive batches of loaves, then the loaves of each batch will resemble those of the preceding batch.

We have in this respect a very close analogy with what takes place in heredity. The egg consists of a certain quantity of germ-plasm and this germ-plasm has certain characteristic peculiarities of its own. In order that it may develop into an adult organism like the parent, it must be subjected to a certain treatment. In the case of a hen's egg undergoing incubation, or of the human fœtus developing in the womb of the mother, we may truthfully say that it has to be baked in an oven at a particular temperature. Only if all the conditions are accurately fulfilled will the egg develop into an organism resembling the parent, and it does so simply because the same causes must always produce the same effects. If you start with identical germ-plasm and expose it to identical conditions during its development, you must get an identical result. The child must resemble the parent. It is, indeed, easy to show by experiment that if you vary the conditions you will get either no result at all or a different one. Up to a certain point, however, the living organism has the power of counter-acting accidental influences, and thereby maintaining its normality of structure. In other words, it is self-regulating, and seems to be always endeavouring to carry out the plan of structure characteristic of the species to which it belongs, so that, if this plan be disturbed, it will, within limits, be restored again by appropriate growth and readjustment. This power of adhering to a predetermined structural plan, in spite of disturbing influences, is one of the most distinctive attributes of living beings, and must on no account be lost sight of in considering the problem of heredity ; but at the same time it is a power that is strictly limited.

A well-known American investigator, Professor Stockard, has shown, in the case of various animals, how abnormalities can be produced by simply lowering the temperature during development. Some years ago the same observer obtained even more surprising results by the use of a simple chemical reagent. He exposed the eggs of the American sea-minnow (*Fundulus*) to the action of magnesium chloride, and found that the young fish tended to develop with a single eye in the middle of the head, instead of one on each side, though the modification was not in all cases complete. Thus we see that it is possible, by the application of a specific chemical stimulus to the egg, to bring about a profound and perfectly definite change in the structure of the organism, though we are still far from knowing why this should be the case.

We also know that the growth of various organs in the animal body is normally controlled by infinitesimal quantities of chemical substances secreted by the ductless glands, such as the thyroid and the pituitary, and circulated in the blood, and that any deficiency or excess of these substances may produce abnormal results. The discovery of these hormones, as they have been termed by Professor Starling, must have a profound influence on our ideas as to the mechanism of heredity. Their significance from this point of view was, I believe, first pointed out by J. T. Cunningham many years ago. It seems at least possible that chemical substances of a like nature may exist in the germ-cells and exercise a profound influence upon their development.

With this possibility in view, let us again examine the egg-cell at the very commencement of its development into a multicellular body, at the moment of its division into the first two daughter cells, and let us concentrate our attention upon the nucleus, which always divides first. As it prepares itself for this important event a number of peculiar bodies called the chromosomes make their appearance, apparently by concentration of previously scattered granules of chromatin substance, so called because of the way in which it can be stained by certain dyes. At the same time a spindle-shaped arrangement of threads becomes manifest and the nuclear membrane disappears, so that there is no longer any sharp division between nucleus and cell-body. The chromosomes, often varying in shape and size amongst themselves, but definite and constant for each kind of organism, arrange themselves across the middle of the spindle. Then each splits into two, and one half moves away from the other and towards the corresponding end of the spindle. We have now two groups of daughter chromosomes, and around each group a new nucleus is constituted. Then the protoplasm of the cell-body divides into two parts, and two complete cells are formed, each with its own nucleus.

The process is really far more complicated than this brief and inadequate description might lead you to suppose, but the essential feature seems always to be the behaviour of the chromosomes. It is very evident that the protoplasm of which they are composed must be of the utmost importance to the organism, and that it is necessary that it should be very accurately divided between the daughter cells every time cell-division takes place. This phenomenon of *mitosis*, as it is termed, is of almost universal occurrence throughout the animal and vegetable kingdoms, not only in the early divisions of the egg-cell, but throughout the entire life of the organism, whenever cell-division takes place. It is clearly a contrivance by which a certain material substance—a particular kind of living protoplasm—is accurately distributed amongst the progeny of a dividing cell. In other words, it is part of the mechanism of inheritance.

Let us now turn aside for a moment and glance very briefly at another and totally different line of evidence, leading to results which confirm and explain in a very remarkable manner those which we have already arrived at. I refer, of course, to the modern experiments in the breeding of plants and animals, undertaken under the influence of what is frequently termed the Mendelian school. It is utterly impossible to do justice to these wonderful experiments in the time at our disposal. I would point out, in the first place, however, that they have led quite independently to the striking conclusion that there must exist in the protoplasm of the germ-cells definite material entities—the so-called Mendelian factors—which are in some way or other responsible for the appearance in the adult organism of special features—the so-called unit characters—capable of being handed on from one generation to another by the process of heredity. Assuming them to be located in the chromosomes, the behaviour of these factors in inheritance, the permutations and combinations of unit characters which arise in cross-breeding, can be adequately explained by the behaviour of the chromosomes actually observed at certain critical periods of the life-cycle.

Take, for example, the colour of the human eye. If a certain factor, or combination of factors, alone be present in the germ-plasm, the eye will be blue or grey, but the addition of another factor may cause it to be brown, and the average results, as regards eye colour, of mating pure blue-eyed and pure brown-eyed individuals can be confidently predicted. The occasional appearance of an extra thumb or finger upon the hand, which is well known to be a heritable character, transmitted with great regularity from parent to child, is again supposed to be due to the occurrence of a corresponding factor in the germ-plasm, and so on with a whole host of characters that have been carefully investigated by means of breeding experiments in recent years. It is important to note that these characters seem to bear no purposeful relation whatever to the well-being of the organism in which they occur. They are often extremely insignificant, and a large proportion of them must undoubtedly be regarded as abnormalities. It is a mere matter of chance whether they happen to be useful, neutral, or injurious.

The investigations of Professor Morgan and his colleagues have gone so far as to demonstrate conclusively, albeit indirectly, not only that the Mendelian factors must be located in the chromosomes of the nucleus, but also that they must be arranged in each chromosome in a perfectly definite manner. These observers have even prepared maps of chromosomes showing the arrangement of the factors in linear series. It is surely one of the most remarkable achievements of modern science that we should be able to point to a particular spot in a particular chromosome of a microscopic germ-cell and say with confidence that there is something just there that is responsible for some particular character, such as the colour of the eye, in the adult organism.

As to the nature of the factors themselves, it seems not unreasonable to conclude that they must consist of definite chemical substances, or, perhaps better, of chemical modifications of living protoplasm, in the form of minute particles too small to be rendered visible by any means yet discovered, but, capable of self-multiplication like other protoplasmic units.

We may further suppose that these factor-forming sub-stances play a part in controlling the development of the organism comparable with that played by the magnesium chloride in the case of the developing embryos of the sea-minnow, or by other chemical substances (hormones) in the normal adult animal. The complex mechanism of mitosis in the division of the cell-nucleus would then appear to be necessary in order to secure the proper distribution of factors throughout the growing body, so that each may reach the particular part that it is destined to influence.

It must be remembered that the occurrence of Mendelian phenomena in heredity depends entirely upon a much more fundamental phenomenon—that of sex—which gives the experimenter the opportunity of crossing two individuals differing as to one or more separately heritable characters, and of observing the numerical proportions of the offspring in which each of these characters makes its appearance.

The phenomenon of sex, as we all know, is a very great mystery, and introduces endless complications into life. Sexual differentiation appears to be nearly as old as the cell itself. The stream of life, almost since it first began to flow, has been a double stream, or, better, a network, in which male and female streamlets unite at more or less frequent intervals to form those temporary overflows which we call individuals. Each streamlet goes its own way for a time, and then joins and exchanges experiences, so to speak, with another. It is just this exchange of experiences that forms the basis of the Men-delian phenomena, and it is not merely the experiences of a single lifetime, but those of many generations that may be thus exchanged.

Perhaps, however, we are getting a little too metaphorical and had better consider in a rather more matter-of-fact manner what actually takes place in the sexual process. The essential feature of this process is always the same—the union of two germ-cells to form a single cell, although this fundamental act is greatly obscured in the higher plants and animals by the endless contrivances which have arisen in the course of evolution, and which serve the ultimate purpose of bringing the germ-cells together. In all the higher animals and plants these germ-cells are sharply differentiated into male and female, spermatozoa or sperm-cells and ova or egg-cells, and, with rare exceptions, the egg-cell cannot even begin to develop until it has united with, or, as we say, been fertilized by, a sperm-cell. This is very literally the union of two branches of the stream of life.

From the point of view of the theory of heredity, the most important thing about this union is the coming together of two sets of chromosomes—paternal and maternal—the one set coming with the spermatozoon from the male parent, and the other with the ovum from the female parent. The maternal and paternal chromosomes bring with them factors that have arisen in some unknown way, probably by chemical changes, in the two ancestral streams of protoplasm which unite in the fertilized egg. Apart altogether from the much-vexed question of the inheritance of " acquired " characters, which we cannot even touch upon this evening, these factors represent certain experiences which the stream of life has gathered in its journey.

Hence the new organism may exhibit certain characters derived from the father and others derived from the mother, a combination of paternal and maternal peculiarities, while the fundamental features of its organization cannot be said to be derived from the one parent more than from the other. It will resemble either parent just in so far as it starts life with the same potentialities, inherent in the germ-plasm as a whole, and in its special factors, and just in so far as it develops under identical conditions. (We must not forget, though the point is not essential to our argument, that the germ-cells may perhaps contain other special factors besides those which have been located in the chromosomes.)

It is a curious fact, and one upon which social reformers and preachers of equality would do well to reflect more seriously, that the characters, the potentialities for good or evil, of a living being should depend so much upon mere chance. A great deal can be done for the welfare of the individual by improving the conditions under which it lives, as every gardener knows, but nothing can altogether counteract the effects of hereditary tendencies. It is worth while to consider a little more fully how it comes about that chance plays such an important part.

With certain exceptions, which do not affect the general proposition, every cell of the living organism contains, as we have already seen, a double set of chromosomes, one set derived from the male and the other from the female parent. The duplication takes place at the time when the two germ-cells come together to form the fertilized egg, but it is counter-acted again by the reduction at another period of the life-cycle ; otherwise the number of chromosomes in each cell would continue to increase in geometrical ratio from generation to generation, which is clearly impossible. It is at these two critical periods that chance steps in and prepares her surprises.

In the first place it seems to be purely a matter of chance what luck the germ-cells have in their mating, what particular ovum is fertilized by what particular spermatozoon, and, owing to the enormous numbers in which ova and sperma-tozoa are produced, the possibilities may be almost infinite. In the second place there are many alternative possibilities with regard to the particular factors which any given germ-cell, male or female, may contain. This depends upon which particular chromosomes happen to remain in the germ-cell after the double number has been halved again. In animals this halving takes place at the time when the germ-cells are ripening, shortly before they are ready to unite in the fertil-ized egg. The maternal and paternal chromosomes in each, differing as regards the factors they contain, pair off during the process of mitosis. The members of each pair then separate, and one of them alone remains in each mature germ-cell. Hence the germ-cells, even of the same individual, come to differ amongst themselves to a practically unlimited extent as regards their factorial constitution. The life of the individual is like a game of cards, in which a very great deal depends upon the shuffling of the pack, and the player has to do the best he can with the hand dealt out to him. He may make a hopeless failure of it, or a great success ; but still the stream of life flows on, ever gathering and combining new experiences, ever forming itself into fleeting individualities and leaving them to perish on its banks as it passes on to fresh attempts at self-expression.

The interest of the Mendelian breeding experiments is so absorbing that it is little wonder if more fundamental aspects of the problem of heredity, to which we have alluded in the earlier part of our lecture, have been largely lost sight of in recent years, while the factorial hypothesis has been hailed by some extremists as the all-sufficient explanation of everything. The characters of the organism may indeed be modified by factors in the germ-plasm, just as the character of a loaf may be modified by putting caraway seeds into the dough ; but the caraway seeds do not explain the loaf, and the Mendelian factors cannot explain the organism as a whole. There is doubtless a good deal of truth in the old saying that life is made up of trifles ; but it is not the whole truth, and the body of a living organism cannot be regarded as merely the sum-total of its unit characters.

Whatever may be their significance from the point of view of the general theory of evolution and heredity, however, there can be but one opinion as to the immense practical importance of the Mendelian investigations. They have already led to the production of many valuable forms of life, more especially plants, that are to all intents and purposes new creations, although their value and novelty may depend merely upon the bringing together of desirable characters in new combinations and the elimination of undesirable features.

Nor are the possibilities of improvement by selective mating confined to our domesticated plants and animals. Hopes are entertained by many enthusiasts, banded together in the interests of what they have thought fit to term the science of eugenics, of effecting vast improvements in the human race itself by the application of Mendelian principles. It does not seem likely, however, or even desirable, that men and women should ever consent to be guided in their choice of mates by purely utilitarian considerations.

There are many objections to any far-reaching schemes of this kind, but it does seem possible, when once the facts of heredity are generally known, that the exercise of an enlightened public opinion and individual choice may result in the elimination from the stream of human life of many heritable characteristics which it is very undesirable to perpetuate. In extreme cases, such as feeble-mindedness and certain forms of insanity, it may even be necessary for the community to protect itself by legislation against the criminal propagation of the unfit.

What is wanted, first and foremost, however, is education, and I trust that you will agree with me that it is education in biology—the science of life—to which we may most hopefully look for the physical and mental improvement of the human race. Men and women must learn to realize their responsibilities towards future generations from the biological point of view, and it is in this direction that the citizens of a great city like Edinburgh can best help, by generously supporting the cause of education and research as represented by your ancient and world-famous University.

Yes, as Dendy suggests towards the end of his lecture, evolution which has worked without interference for millions of years, may now be controlled to some extent by man. So before we turn to the complicated but fascinating story of genetics, or the machinery of heredity, let us glance at the history of evolutionary theory.

As Sir J. G. Frazer pointed out, the creation myths of primitive peoples are of two kinds : tillers of the soil, dependent on the yearly miracle of seed-time and harvest, feel that man and all the other animals were formed from the dust by some Creator who breathed into their clay the breath of life. Hunters and herders, on the other hand, feeling the close relationship of vegetable and animal, and of animal and man, tend to believe in the changing of one form of life into another, so that a whole tribe may claim descent from its totem animal ancestor. These two lines of thought can be traced all through history until Darwin, with his theory of natural selection, supported by a vast array of facts, made the case for evolution overwhelmingly strong. Many thinkers had believed in evolution before him, but they could not find any satisfactory explanation of how it had worked. Aristotle thought that all animals might have developed from an egg or a grub. St. Augustine said that " in the grain itself there were invisible all things simultaneously which were in time to grow into the tree ". But the first serious attempt to explain how evolution might have worked was that by Lamarck early in the nineteenth century. Lamarck's theory depended on the inheritance of acquired characters ; organs which were not used tended to become smaller and weaker, and offspring would be born with these organs already weakened, while " the frequent use of any organ, when confirmed by habit, increases the functions of that organ, leads to its development and endows it with a size and power that it does not possess in animals which exercise it less ". Those are Lamarck's own words and he supported his view with a reference to the webbed feet of swimming birds, but his claims carried little conviction.

In 1830 the way was prepared for Darwin by Lyell's *Principles of Geology* which showed that the surface of the earth had been sculptured by the slow action of everyday forces working through millions of years. A year later Darwin set off on the voyage of the *Beagle*, which lasted five years, and during which he accumulated an enormous mass of notes and geological specimens. After his return to England he continued to classify and arrange and brood over his material, and finally in 1859 *The Origin of Species* was published. Alfred Russell Wallace had independently come to the same conclusions as Darwin at about the same time, and instead of any unseemly squabble as to priority, a joint paper was given to the Linnean Society in 1858. *How does " the survival of the fittest " lead to the origin of new species ?* Here are some selections from Darwin's book which will answer this question. So much has falsely been called " Darwinism " that it is just as well to have a statement of Darwin's views in his own words.

CHARLES DARWIN

NATURAL SELECTION IN EVOLUTION

NO one ought to feel surprised at much remaining as yet unexplained in regard to the origin of species and varieties, if he make due allowance for our profound ignorance in regard to the mutual relations of the many beings which live around us. Who can explain why one species ranges widely and is very numerous, and why another allied species has a narrow range and is rare ? Yet these relations are of the highest importance, for they determine the present welfare and, as I believe, the future success and modification of every inhabitant of this world. Still less do we know of the mutual relations of the innumerable inhabitants of the world during the many past geological epochs in its history. Although much remains obscure, and will long remain obscure, I can entertain no doubt, after the most deliberate study and dispassionate judgment of which I am capable, that the view which most naturalists until recently entertained, and which I formerly entertained—namely, that each species has been independently created—is erroneous. I am fully convinced that species are not immutable ; but that those belonging to what are called the same genera are lineal descendants of some other and generally extinct species, in the same manner as the acknowledged varieties of any one species are the descendants of that species. Furthermore I am convinced that Natural Selection has been the most important, but not the exclusive, means of modification.

EFFECTS OF HABIT; CORRELATED VARIATION; INHERITANCE

Changed habits produce an inherited effect, as in the period of the flowering of plants when transported from one climate to another. With animals the increased use or disuse of parts has had a more marked influence ; thus, I find in the domestic duck that the bones of the wing weigh less and the bones of the leg more, in proportion to the whole skeleton, than do the same bones in the wild duck ; and this change may be safely attributed to the domestic duck flying much less, and walking more, than its wild parents. The great and inherited development of the udders in cows and goats in countries where they are habitually milked, in comparison with these organs in other countries, is probably another instance of the effects of use. Not one of our domestic animals can be named which has not in some country drooping ears ; and the view which has been suggested that the drooping is due to disuse of the muscles of the ear, from the animals being seldom much alarmed, seems probable.

Many laws regulate variation, some few of which can be dimly seen, and will hereafter be briefly discussed. I will here only allude to what may be called correlated variation. Important changes in the embryo or larva will probably entail changes in the mature animal. In monstrosities, the correlations between quite distinct parts are very curious ; and many instances are given in Isidore St. Hilaire's great work on this subject. Breeders believe that long limbs are almost always accompanied by an elongated head. Some instances of correlation are quite whimsical : thus, cats which are entirely white and have blue eyes are generally deaf; but it has been lately stated by Mr. Tate that this is confined to the males. Colour and constitutional peculiarities go together, of which many remarkable cases could be given amongst animals and plants.

GEOMETRICAL RATIO OF INCREASE

A struggle for existence inevitably follows from the high rate at which all organic beings tend to increase. Every being which during its natural lifetime produces several eggs or seeds must suffer destruction during some period of its life, and during some season or occasional year, otherwise, on the principle of geometrical increase, its numbers would quickly become so inordinately great that no country could support the product. Hence, as more individuals are produced than can possibly survive, there must in every case be a struggle for existence, either one individual with another of the same species, or with the individuals of distinct species, or with the physical conditions of life. It is the doctrine of Malthus applied with manifold force to the whole animal and vegetable kingdoms ; for in this case there can be no artificial increase of food and no prudential restraint from marriage. Although some species may be now increasing, more or less rapidly, in numbers, all cannot do so, for the world would not hold them.

THE SURVIVAL OF THE FITTEST

How will the struggle for existence, briefly discussed in the last chapter, act in regard to variation ? Can the principle of selection, which, we have seen, is so potent in the hands of man, apply under Nature ? I think we shall see that it can act most efficiently. Let the endless number of slight variations and individual differences occurring in our domestic productions, and, in a lesser degree, in those under Nature, be borne in mind ; as well as the strength of the hereditary tendency. Under domestication, it may be truly said that the whole organization becomes in some degree plastic. But the variability which we almost universally meet within our domestic productions is not directly produced, as Hooker and Asa Gray have well remarked, by man ; he can neither originate varieties nor prevent their occurrence ; he can only preserve and accumulate such as do occur. Unintentionally he exposes organic beings to new and changing conditions of life, and variability ensues ; but similar changes of conditions might and do occur under Nature. Let it also be borne in mind how infinitely complex and close-fitting are the mutual relations of all organic beings to each other and to their physical conditions of life ; and consequently what infinitely varied diversities of structure might be of use to each being under changing conditions of life. Can it, then, be thought improbable, seeing that variations useful to man have undoubtedly occurred, that other variations useful in some way to each being in the great and complex battle of life, should occur in the course of many successive generations? If such do occur, can we doubt (remembering that many more individuals are born than can possibly survive) that individuals having any advantage, however slight, over others would have the best chance of surviving and of procreating their kind? On the other hand, we may feel sure that any variation in the least degree injurious would be rigidly destroyed. This preservation of favourable individual differences and variations, and the destruction of those which are injurious, I have called Natural Selection, or the Survival of the Fittest. Variations neither useful nor injurious would not be affected by natural selection, and would be left either a fluctuating element, as perhaps we see in certain polymorphic species, or would ultimately become fixed, owing to the nature of the organism and the nature of the conditions.

We shall best understand the probable course of natural selection by taking the case of a country undergoing some slight physical change, for instance, of climate. The proportional numbers of its inhabitants will almost immediately undergo a change, and some species will probably become extinct. We may conclude, from what we have seen of the intimate and complex manner in which the inhabitants of each country are bound together, that any change in the numerical proportions of the inhabitants, independently of the change of climate itself, would seriously affect the others. If the country were open on its borders, new forms would certainly immigrate, and this would likewise seriously disturb the relations of some of the former inhabitants. Let it be remembered how powerful the influence of a single introduced tree or mammal has been shown to be. But in the case of an island, or of a country partly surrounded by barriers, into which new and better adapted forms could not freely enter, we should then have places in the economy of Nature which would assuredly be better filled up, if some of the original inhabitants were in some manner modified ; for, had the area been open to immigration, these same places would have been seized on by intruders. In such cases, slight modifications which in any way favoured the individuals of any species by better adapting them to their altered conditions would tend to be preserved ; and natural selection would have free scope for the work of improvement.

We have good reason to believe, as shown in the first section, that changes in the conditions of life give a tendency to increased variability, and in the foregoing cases the conditions have changed, and this would manifestly be favourable to natural selection, by affording a better chance of the occurrence of profitable variations. Unless such occur, natural selection can do nothing. Under the term of " variations ", it must never be forgotten that mere individual differences are included. As man can produce a great result with his domestic animals and plants by adding up in any given direction individual differences, so could natural selection, but far more easily from having incomparably longer time for action. Nor do I believe that any great physical change, as of climate, or any unusual degree of isolation to check immigration, is necessary in order that new and unoccupied places should be left for natural selection to fill up by improving some of the varying inhabitants. For, as all the inhabitants of each country are struggling together with nicely balanced forces, extremely slight modification in the structure or habits of one species would often give it an advantage over others ; and still further modifications of the same kind would often still further increase the advantage, as long as the species continued under the same conditions of life and profited by similar means of subsistence and defence. No country can be named in which all the native inhabitants are now as perfectly adapted to each other and to the physical conditions under which they live, that none of them could be still better adapted or improved ; for in all countries the natives have been so far conquered by naturalized productions, that they have allowed some foreigners to take firm possession of the land. And as foreigners have thus in every country beaten some of the natives, we may safely conclude that the natives might have been modified with advantage, so as to have better resisted the intruders.

As man can produce, and certainly has produced, a great result by his methodical and unconscious means of selection, what may not natural selection effect ? Man can act only on external and visible characters : Nature, if I may be allowed to personify the natural preservation or survival of the fittest, cares nothing for appearances, except in so far as they are useful to any being. She can act on every internal organ, on every shade of constitutional difference, on the whole machinery of life. Man selects only for his own good : Nature only for that of the being which she tends. Every selected character is fully exercised by her, as is implied by the fact of their selection. Man keeps the natives of many climates in the same country ; he seldom exercises each selected character in some peculiar and fitting manner ; he feeds a long and a short-beaked pigeon on the same food ; he does not exercise a long-backed or long-legged quadruped in any peculiar manner ; he exposes sheep with long and short wool to the same climate. He does not allow the most vigorous males to struggle for the females. He does not rigidly destroy all inferior animals, but protects during each varying season, as far as lies in his power, all his productions. He often begins his selection by some half-monstrous form ; or at least by some modification prominent enough to catch the eye or to be plainly useful to him. Under Nature, the slightest differences of structure or constitution may well turn the nicely balanced scale in the struggle for life, and so be preserved. How fleeting are the wishes and efforts of man ! how short his time ! and consequently how poor will be his results, compared with those accumulated by Nature during whole geological periods ! Can we wonder, then, that Nature's productions should be far "truer" in character than man's productions; that they should be infinitely better adapted to the most complex conditions of life, and should plainly bear the stamp of far higher workmanship?

It may metaphorically be said that natural selection is daily and hourly scrutinizing, throughout the world, the slightest variations; rejecting those that are bad, preserving and adding up all that are good; silently and insensibly working whenever and wherever opportunity offers, at the improvement of each organic being in relation to its organic and inorganic conditions of life. We see nothing of these slow changes in progress, until the hand of time has marked the lapse of ages, and then so imperfect is our view into long-past geological ages, that we see only that the forms of life are now different from what they formerly were.

In order that any great amount of modification should be effected in a species, a variety when once formed must again, perhaps after a long interval of time, vary and present individual differences of the same favourable nature as before; and these must be again preserved, and so onwards step by step. Seeing that individual differences of the same kind perpetually recur, this can hardly be considered as an unwarrantable assumption. But whether it is true, we can judge only by seeing how far the hypothesis accords with and explains the general phenomena of Nature. On the other hand, the ordinary belief that the amount of possible variation is a strictly limited quantity is likewise a simple assumption.

Although natural selection can act only through and for the good of each being, yet characters and structures, which we are apt to consider as of very trifling importance, may thus be acted on. When we see leaf-eating insects green, and bark-feeders mottled-grey; the alpine ptarmigan white in winter, and red-grouse the colour of heather, we must believe that these tints are of service to these birds and insects in preserving them from danger. Grouse, if not destroyed at some period of their lives, would increase in countless numbers; they are known to suffer largely from birds of prey ; and hawks are guided by eyesight to their prey—so much so, that on parts of the Continent persons are warned not to keep white pigeons, as being the most liable to destruction. Hence natural selection might be effective in giving the proper colour to each kind of grouse, and in keeping that colour, when once acquired, true and constant. Nor ought we to think that the occasional destruction of an animal of any particular colour would produce little effect : we should remember how essential it is in a flock of white sheep to destroy a lamb with the faintest trace of black. We have seen how the colour of the hogs, which feed on the " paint-root " in Virginia, determines whether they shall live or die. In plants, the down on the fruit and the colour of the flesh are considered by botanists as characters of the most trifling importance : yet we hear from an excellent horticulturist, Downing, that in the United States smooth-skinned fruits suffer far more from a beetle, a curculio, than those with down ; that purple plums suffer far more from a certain disease than yellow plums ; whereas another disease attacks yellow-fleshed peaches far more than those with other coloured flesh. If, with all the aids of art, these slight differences make a great difference in cultivating the several varieties, assuredly, in a state of Nature, where the trees would have to struggle with other trees and with a host of enemies, such differences would effectually settle which variety, whether a smooth or downy, a yellow or purple-fleshed fruit, should succeed. . . .

Sexual Selection

Inasmuch as peculiarities often appear under domestication in one sex and become hereditarily attached to that sex, so no doubt it will be under Nature. Thus it is rendered possible for the two sexes to be modified through natural selection in relation to different habits of life, as is sometimes the case ; or for one sex to be modified in relation to the other sex, as commonly occurs. This leads me to say a few words

on what I have called Sexual Selection. This form of selection depends, not on a struggle for existence in relation to
other organic beings or to external conditions, but on a struggle
between the individuals of one sex, generally the males, for
the possession of the other sex. The result is not death to the
unsuccessful competitor, but few or no offspring. Sexual
selection is, therefore, less rigorous than natural selection.
Generally, the most vigorous males, those which are best
fitted for their places in Nature, will leave most progeny. But
in many cases, victory depends not so much on the general
vigour, as on having special weapons, confined to the male
sex. A hornless stag or spurless cock would have a poor
chance of leaving numerous offspring. Sexual selection, by
always allowing the victor to breed, might surely give indomit-
able courage, length to the spur, and strength to the wing to
strike in the spurred leg, in nearly the same manner as does
the brutal cock-fighter by the careful selection of his best
cocks. How low in the scale of Nature the law of battle
descends, I know not ; male alligators have been described
as fighting, bellowing, and whirling around like Indians in
a war-dance, for the possession of the females ; male salmon
have been observed fighting all day long ; male stag-beetles
sometimes bear wounds from the huge mandibles of other
males ; the males of certain hymenopterous insects have
been frequently seen by that inimitable observer, M. Fabre,
fighting for a particular female, who sits by, an apparently
unconcerned beholder of the struggle, and then retires with
the conqueror. The war is, perhaps, severest between the
males of polygamous animals, and these seem oftenest provided with special weapons. The males of carnivorous animals
are already well armed ; though to them and to others special
means of defence may be given through means of sexual
selection, as the mane of the lion, and the hooked jaw to the
male salmon ; for the shield may be as important for victory
as the sword or spear.

Amongst birds, the contest is often of a more peaceful

character.   All those who have attended to the subject believe that there is the severest rivalry between the males of many species to attract, by singing, the females.   The rock-thrush of Guiana, birds of paradise, and some others, congregate ; and successive males display with the most elaborate care, and show off in the best manner, their gorgeous plumage ; they likewise perform strange antics before the females, which, standing by as spectators, at last choose the most attractive partner.   Those who have closely attended to birds in confinement well know that they often take individual preferences and dislikes : thus Sir R. Heron has described how a pied peacock was eminently attractive to all his hen birds.   I cannot here enter on the necessary details ; but if man can in a short time give beauty and an elegant carriage to his bantams, according to his standard of beauty, I can see no good reason to doubt that female birds, by selecting, during thousands of generations, the most melodious or beautiful males, according to their standard of beauty, might produce a marked effect.   Some well-known laws, with respect to the plumage of male and female birds in comparison with the plumage of the young, can partly be explained through the action of sexual selection of variations occurring at different ages, and transmitted to the males alone or to both sexes at corresponding ages ; but I have not space here to enter on this subject.

Thus it is, as I believe, that when the males and females of any animal have the same general habits of life, but differ in structure, colour, or ornament, such differences have been mainly caused by sexual selection : that is, by individual males having had, in successive generations, some slight advantage over other males, in their weapons, means of defence, or charms, which they have transmitted to their male offspring alone.   Yet I would not wish to attribute all sexual differences to this agency : for we see in our domestic animals peculiarities arising and becoming attached to the male sex, which apparently have not been augmented through selection by

man. The tuft of hair on the breast of the wild turkey-cock
cannot be of any use, and it is doubtful whether it can be
ornamental in the eyes of the female bird ; indeed, had the
tuft appeared under domestication it would have been called
a monstrosity.

THE ACTION OF NATURAL SELECTION

In order to make it clear how, as I believe, natural selection
acts, I must beg permission to give one or two imaginary
illustrations. Let us take the case of a wolf, which preys on
various animals, securing some by craft, some by strength, and
some by fleetness ; and let us suppose that the fleetest prey, a
deer for instance, had from any change in the country increased
in numbers, or that other prey had decreased in numbers,
during that season of the year when the wolf was hardest
pressed for food. Under such circumstances the swiftest and
the slimmest wolves would have the best chance of surviving
and so be preserved or selected, provided always that they
retained strength to master their prey at this or some other
period of the year when they were compelled to prey on other
animals. I can see no more reason to doubt that this would
be the result than that man should be able to improve the
fleetness of the greyhound by careful and methodical selection, or by that kind of unconscious selection which follows
from each man trying to keep the best dogs without any
thought of modifying the breed. I may add, that, according
to Mr. Pierce, there are two varieties of the wolf inhabiting
the Catskill Mountains, in the United States, one with a light
greyhound-like form, which pursues deer, and the other more
bulky, with shorter legs, which more frequently attacks the
shepherd's flocks. . . .

SUMMARY

If under changing conditions of life organic beings present
individual differences in almost every part of their structure,

and this cannot be disputed ; if there be, owing to their geometrical rate of increase, a severe struggle for life at some age, season, or year, and this certainly cannot be disputed ; then, considering the infinite complexity of the relations of all organic beings to each other and to their conditions of life, causing an infinite diversity in structure, constitution, and habits, to be advantageous to them, it would be a most extra-ordinary fact if no variations had ever occurred useful to each being's own welfare, in the same manner as so many variations have occurred useful to man. But if variations useful to any organic being ever do occur, assuredly individuals thus char-acterized will have the best chance of being preserved in the struggle for life ; and from the strong principle of inheritance, these will tend to produce offspring similarly characterized. This principle of preservation, or the survival of the fittest, I have called Natural Selection. It leads to the improvement of each creature in relation to its organic and inorganic conditions of life ; and consequently, in most cases, to what must be regarded as an advance in organization. Nevertheless, low and simple forms will long endure if well fitted for their simple conditions of life.

Natural selection, on the principle of qualities being in-herited at corresponding ages, can modify the egg, seed, or young, as easily as the adult. Amongst many animals, sexual selection will have given its aid to ordinary selection, by assuring to the most vigorous and best adapted males the greatest number of offspring. Sexual selection will also give characters useful to the males alone in their struggles or rivalry with other males ; and these characters will be transmitted to one sex or to both sexes, according to the form of inheritance which prevails.

Whether natural selection has really thus acted in adapting the various forms of life to their several conditions and stations must be judged by the general tenor and balance of evidence given in the following chapters. But we have already seen how it entails extinction ; and how largely extinction has acted in the world's history, geology plainly declares. Natural selection, also, leads to divergence of character; for, the more organic beings diverge in structure, habits, and constitution, by so much the more can a large number be supported on the area, of which we see proof by looking to the inhabitants of any small spot, and to the productions naturalized in foreign lands. Therefore, during the modification of the descendants of any one species, and during the incessant struggle of all species to increase in numbers, the more diversified the descendants become, the better will be their chance of success in the battle for life. Thus the small differences distinguishing varieties of the same species steadily tend to increase till they equal the greater differences between species of the same genus, or even of distinct genera. . .

The affinities of all the beings of the same class have sometimes been represented by a great tree. I believe this simile largely speaks the truth. The green and budding twigs may represent existing species; and those produced during former years may represent the long succession of extinct species. At each period of growth all the growing twigs have tried to branch out on all sides, and to overtop and kill the surrounding twigs and branches, in the same manner as species and groups of species have at all times overmastered other species in the great battle for life. The limbs divided into great branches, and these into lesser and lesser branches, were themselves once, when the tree was young, budding twigs; and this connexion of the former and present buds by ramifying branches may well represent the classification of all extinct and living species in groups subordinate to groups. Of the many twigs which flourished when the tree was a mere bush, only two or three, now grown into great branches, yet survive and bear the other branches; so with the species which lived during long-past geological periods, very few have left living and modified descendants. From the first growth of the tree, many a limb and branch has decayed and dropped off; and these fallen branches of various sizes may represent those whole orders, families, and genera which have now no living representatives, and which are known to us only in a fossil state. As we here and there see a thin straggling branch springing from a fork low down in a tree, and which by some chance has been favoured and is still alive on its summit, so we occasionally see an animal like the Ornithorhynchus or Lepidosiren, which in some small degree connects by its affinities two large branches of life, and which has apparently been saved from fatal competition by having inhabited a protected station. As buds give rise by growth to fresh buds, and these, if vigorous, branch out and overtop on all sides many a feebler branch, so by generation, I believe, it has been with the great Tree of Life, which fills with its dead and broken branches the crust of the earth, and covers the surface with its ever-branching and beautiful ramifications.

In the century which has passed since Darwin brooded over his fossils in Down House the study of evolution has been carried on with intense eagerness in every civilized country. The flood of new light has altered our ideas of how the machinery works, but nothing has been, or is ever likely to be, discovered to alter the scientists' view that evolution is a fact. *What is man's place in the evolutionary scheme?* For an answer to this question we turn again to Professor Dendy, who takes Darwin's symbol of a great tree of life, and shows how Man has climbed it.

ARTHUR DENDY

BIOLOGICAL FOUNDATIONS OF SOCIETY

THE sharp distinction commonly drawn between mankind and the lower animals, rooted, so far as the western world is concerned, in the Biblical fable of the creation, has, from the evolutionary standpoint at any rate, long ceased to be tenable as a logical antithesis. The light cast upon the origin of the human race by biological research has penetrated many dark places and left no shadow of justification for the old-fashioned dogmas of religious orthodoxy with regard to this important question. Indeed, orthodoxy itself has of late years found it necessary to abandon many of its own strongholds and seek a place more closely in touch with the advancing army of science.

As my entire argument rests upon the evolutionary hypothesis and its application to mankind, I may be excused for recapitulating very briefly some of the evidence by which our belief in that hypothesis is supported, though I am somewhat loath to take this course for fear of giving, in the limited space at our disposal, an altogether inadequate impression of the overwhelming strength of the case.

The human body, alike in its structure, functions, and development, exhibits the most unmistakable traces of its pre-human ancestry, and of the common origin of mankind and the lower animals. Anatomically considered, and apart from the brain, the body of a man is actually less highly organized, a less complex and less specialized piece of mechanism, than the bodies of many other vertebrates. The limbs in particular, in their typical five-fingered or five-toed structure, have retained a primitive character, and are far less advanced in evolution than the wings of a bird, or even the legs of a horse. The bird, as a flying animal, has an immense advantage over the man, and it is only at the present day, by the invention of flying machines, that man is beginning to make good this deficiency in his bodily organization.

If we examine the wing of a bird, or the fore-limb of a horse, however, in comparison with the human arm, we shall find that, in spite of their great difference in external appearance and mode of action, they all exhibit the same fundamental plan of structure. All can be derived from the primitive penta-dactyl or five-fingered type characteristic of the whole of the air-breathing vertebrates. In all, we recognize the arm, supported by the humerus, the forearm with the radius and ulna, the wrist with its carpal bones, and the hand with its metacarpals and phalanges. The differences are due simply to exaggeration, suppression, or fusion of these primitive constituents, and the only rational explanation of the fact that such different organs as arms, legs, and wings exhibit essentially the same structural plan is that that plan has been inherited from some common ancestor, and adapted to the special requirements of each particular case.

So also with all the other organs of the human body, we find their counterpart, more or less exactly, in the bodies of other vertebrates, and it is only as regards details that we see any great variations in structure. There is, in short, one fundamental plan of organization common to all the great group Vertebrata ; a plan which comprises the dorsally situated central nervous system, differentiated into brain and spinal cord ; the cranium or brain case surrounding the brain ; the vertebral column, based upon the embryonic notochord and surrounding the spinal cord ; the ventrally situated heart and the respiratory pharynx. These are all characteristic vertebrate features, but the vertebrate type itself is but a modification of a still more fundamental plan of organization which the vertebrates share with some of the invertebrate classes. These latter again possess many features in common, while all of them, vertebrates and invertebrates alike, are, on microscopic analysis, found to be built up of essentially similar structural units—those microscopical, nucleated masses of protoplasm known as cells.

Amongst the Vertebrata we can distinguish five great classes, the Fishes, the Amphibians, the Reptiles, the Birds, and the Mammals, which, at the present day, are fairly sharply marked off from one another, although, when we trace their history backwards through geological time, we find them losing their distinctive features and gradually merging into one another.

The Fishes (amongst which we may, for the sake of simplicity, still include the lampreys) are undoubtedly the most primitive of these five classes, and in large measure represent the ancestral condition in which all the vertebrates originated. They are characterized above all by their thorough adaptation to an aquatic mode of life, entailing the possession of gills and branchial respiration throughout life, with corresponding arrangement of the heart and blood-vessels, while their limbs retain the primitive form of fins, and have not acquired that pentadactyl structure which we have seen to be so characteristic of the air-breathing groups.

Next to the Fishes come the Amphibians, which still retain many fish-like features. Many of them, such as the common newts of our ponds and ditches, are still well adapted to an aquatic life, and some of them keep their gills as functioning organs of respiration throughout the whole of their existence. All of them, however, have developed lungs as outgrowths of the œsophagus, comparable to the swim-bladder of certain fishes, but highly vascularized, and thus forming organs which are capable of breathing air directly. The arrangement of the heart and great blood-vessels becomes modified accordingly, while the primitive fish fin has been transformed into the five-toed limb adapted for locomotion on dry land. Even the frogs and toads, which, in the adult condition, depart so widely from the fish-like form and habit, pass through a very typical fish-like condition in their development—the active, free-living tadpole.

The Reptiles differ from the Amphibians in many important respects, but especially in their more complete adaptation to a terrestrial, air-breathing mode of life. At no period of their life-history do they possess gills ; the lungs are the sole organs of respiration in the adult, and there is a further advance towards complete separation of the pulmonary from the systematic circulation. The free-living, fish-like larval stage has been completely suppressed, and the young animal, in its earlier stages, develops within an egg-shell, where it is further protected by the sac-like amnion and nourished at the expense of the yolk stored up in the egg. The ancestral gills are replaced functionally, up to the time of hatching, by a new respiratory organ, the allantois, developed as a vascular outgrowth from the hinder part of the alimentary canal; but, strange to say, the gill slits in the neck still appear in the embryo, although they have entirely lost their original function.

From the reptilian stock two divergent lines of evolution have originated, culminating in the Birds on the one hand and the Mammals on the other. The former have become highly specialized for a life of great activity in the air. In adaptation to this their fore-limbs have been converted into wings, but their most distinctive peculiarity is the replacement of the reptilian scales, over the greater part of the body, by a covering of feathers, which serve both in the formation of the organs of flight and as a means of maintaining the characteristically high temperature of the body. The separation of the pulmonary and systematic circulations has been fully accomplished by the complete division of the heart into right and left sides which no longer communicate with one another. In their egg-laying habit, however, and in the early development of the young animal, they still agree very closely with their reptilian forefathers.

The mammals, on the other hand, have advanced along a very different path. Instead of feathers they have acquired a covering of hair as a protection against excessive changes of temperature, and this is perhaps their most obvious distinguishing feature, though the name of the group is derived from their unique habit of suckling the young. The great majority of mammals, moreover, differ from the ancestral reptiles in that they produce very minute eggs, without shell or yolk, which, for a longer or shorter period, develop within the womb of the mother; the young animal or fœtus, up to the time of birth, being attached by a special organ, the placenta, to the parent. This very characteristic structure is formed mainly from the allantois, which has now taken on a nutritive as well as a respiratory function, absorbing nourishment by diffusion from the blood of the parent and conveying it to the fœtus through the allantoic circulation.

Certain very primitive types of mammal, the Echidna or spiny ant-eater and the Ornithorhynchus or duck-billed platypus, have survived to the present day in Australia and some of the adjacent islands, and in these, although they possess a hairy covering and suckle their young in a very primitive fashion, many of the reptilian features that have been lost in the more typical members of the class still persist. Thus they lay large and heavily yolked eggs provided with egg-shells, within which the early stages of development are passed outside the body of the parent, and there is, of course, no placenta.

Nowhere else amongst existing vertebrates do we meet with such an admirable example of a connecting link between two great classes as that afforded by these curious survivals, but we have only to search the records of the rocks to find parallel cases amongst the vertebrates of the past. The celebrated Archæopteryx of the Jurassic period, whose fossil remains occur in the lithographic slates of Solenhofen in Bavaria, has long been known as a connecting link between the reptiles and the birds, occupying a position exactly analogous to that of Echidna and Ornithorhynchus between the reptiles and the mammals. In fact the geological evidence, now almost embarrassing in its abundance, entirely supports the conclusions derived from the comparative study of recent vertebrates as to the main lines of evolution within the group.

The order of appearance of the five great classes in geological time harmonizes exactly with the theory that each great advance in organization arose by modification of a pre-existing type ; the fishes giving rise to amphibians, the amphibians to reptiles, and the reptiles to birds and mammals.

The origin of the vertebrate phylum as a whole is far less easy to trace—largely, no doubt, because it took place at such an extremely remote period of the earth's history—and it is not necessary for us to attempt to solve this problem. There can be no doubt, however, that the primitive vertebrates were derived from invertebrate ancestors and that the chain of life has been uninterrupted from the time of the first appearance of the simplest organisms to the present day.

There is one great law of progressive evolution which we may consider briefly at this preliminary stage of our inquiry, though we shall have to refer to it again later on. Each new branch of the evolutionary tree, marked by some characteristic modification of structure, arises, of course, from some parent branch, but it seldom or never springs from near the apex. In other words, it is not the more specialized members of an ancestral group that give rise to fresh outbursts of evolutionary vigour, but more lowly organized and primitive representatives, which have retained a high degree of plasticity and are able to adapt themselves to new environmental conditions. This great principle is very clearly illustrated in the case of the five classes of vertebrate animals. Comparative anatomy and palæontology concur in teaching us that the amphibians arose from fishes of primitive type and not from the highly specialized forms that dominate the seas to-day ; that the reptiles arose from primitive amphibians and not from frogs or toads ; and that birds and mammals in their turn sprang from primitive reptiles. Each great group, after giving off one or more vigorous offshoots in this manner, seems to exhaust itself in highly specialized experiments, which lead in the course of time to extinction.

This is nowhere better seen than in the history of the reptiles, which, after giving off the main stems of the birds and mammals about the commencement of the secondary period of the earth's history, branched out into numerous lines of descent which became highly specialized in adaptation to a great variety of conditions ; crawling or walking on the land, swimming in the sea, or even flying in the air, and often attaining gigantic dimensions. Most of these lines are now

extinct and their place has been taken by the mammalian and
avian descendants of the primitive reptilian stock.

The same great principle is no less clearly seen in the
history of mankind in relation to the remainder of the great
mammalian class to which man belongs. We have already had
occasion to notice that his bodily structure exhibits primitive
features, especially as regards the limbs. In this respect he
and the other members of the order Primates are far less
specialized than most of the mammals, and it was the retention
of this primitive, unspecialized limb structure that rendered
possible the adaptation of the arms as prehensile organs, and
the gradual assumption of the erect attitude and bipedal
method of locomotion that followed upon an arboreal mode of
life. It was this change of habit, and especially the increasing
use of the hands in fashioning and grasping tools, that led, by
way of greatly increased and diversified experience, to that
extraordinary brain development that distinguishes mankind
from all the lower animals and marks the commencement of a
new evolutionary era.

Whatever view we may take as to the exact point at which
the human stock branched off from the line of descent of the
other primates—whether we regard man as a glorified ape or
trace his direct ancestry further back to some more primitive
lemuroid type such as tarsius, makes little difference to our
present argument, and the experts in anthropology may be
safely left to settle their controversies amongst themselves.
For us it is sufficient to insist upon the unity of mankind with
the rest of the animal kingdom. If further proof of this unity
be needed it is to be found in the mode of development of the
individual man, and it is perhaps worth while to consider
briefly this aspect of the question. . . .

One of the most important generalizations as yet reached
by biological science is that known as the Law of Recapitulation, which states that each individual organism, whether
animal or plant, in its development from the egg, passes
through a series of stages which more or less closely resemble

the stages through which its ancestors passed, as adult organisms, in the course of their evolution. The ontogeny or individual life-history repeats the phylogeny or history of the race. We have already had occasion to refer to the tadpole stage in the life-history of the frog as reproducing, with considerable exactitude, the structure of the fish-like ancestor of the amphibia. There is no tadpole stage in man because the changes that have taken place in his method of feeding and protecting the young have rendered a free-living larval form unnecessary and even impossible. But if we examine a human fœtus while it is still sufficiently young we shall find a series of transverse grooves in the neck which are unmistakably equivalent to the gill-slits of the tadpole, and clearly point to some extremely remote fish-like condition. At this stage the human fœtus has so much in common with those of other mammals that it would take an experienced embryologist to tell the difference between them, and, apart from the presence of the placenta, it bears an almost equally close resemblance to the embryos of the reptiles and birds. As we trace the development of all these embryos on towards the adult condition we find the differences between them becoming gradually more and more strongly marked as the distinguishing features of the groups to which they belong make their appearance ; but if we trace it backwards we shall find that they all converge and meet, as it were, in a common starting-point—the egg or ovum—for however much the eggs of the several groups may differ from one another in secondary features, such as the presence or absence of an egg-shell and the amount of food yolk they contain, each is essentially a single cell, a single nucleated mass of protoplasm. This egg may be interpreted as representing a unicellular ancestor, from which all the great multicellular groups of the animal kingdom have originated, while other descendants from the same primitive stock, of a more conservative character, have retained the single-cell condition to the present day and constitute the existing Protozoa.

Thus, while the study of comparative anatomy teaches that, organ for organ, even to comparatively small details, the human body shares a common plan of structure with other mammals, and, to a less extent only, with the lower vertebrates also, the study of embryology affords still more conclusive evidence as to man's place in the animal kingdom and his pre-human ancestry. So far as it goes—and it is to be regretted that as yet it goes only for a comparatively short distance—the geological history of mankind entirely supports the same conclusions. Whatever may be the exact period assigned by geologists as that in which the ancestors of the human race first acquired the right to be regarded as men, there can be no doubt that the human stock represents the most recent off-shoot of any importance from the great vertebrate phylum, and in spite of the present deficiency of fossil remains we may fairly hope for a somewhat more satisfactory series of connecting links in the near future. We shall, however, probably never have anything like a complete chain, because, at the commencement of his evolution as a distinct branch, primitive man had not yet become a dominant type, and doubtless existed in very small numbers and for the most part under conditions which rendered it unlikely that his remains would be preserved for the edification of future generations.

Briefly as we have been obliged to deal with the evidence, we have perhaps said enough to justify us in the assumption that mankind, looked at as a part of the animal kingdom, must be subject to the same laws of evolutionary progress and retrogression as the lower animals. From this point of view the distinction between Nature and man, between natural and artificial products and processes, however convenient it may be, cannot be justified, any more than a corresponding distinction between Nature and insect, natural and of insect origin. From other points of view, no doubt, we see things differently, and the evolution of man's mental, if not his moral qualities, goes far towards justifying us in placing him on a pedestal and according him special treatment. To this aspect of the question we shall return later on. Our immediate object will be to investigate the phenomena of evolution for the animal kingdom as a whole, with a view to determining the conditions that are conducive to well-being and progress and distinguishing them from those that tend in the opposite direction. In order to accomplish this object we must go back a very long way.

The human mind appears to be ever attempting to arrange the data with which it works in sharply defined categories, and undoubtedly this proceeding greatly facilitates the mental operations of the average man. Thus it seems only natural to draw a sharp distinction between the living and the not-living, and to assume that as protoplasmic organisms cannot always have existed on the earth they must have made their appearance there at some definite time. The doctrine of the creation of the world, as enunciated in the book of Genesis, the theory of the spontaneous generation of living things by the sudden conversion of inanimate matter into organisms such as exist at the present day, and the theory of the population of the earth by immigrant germs of life from some other planet, all arose in response to the demand of the human mind for some definite beginning. The inconsistency of this demand is at length recognized by the student of evolution. He realizes the absurdity of seeking a fixed starting-point for the evolutionary process and a definite commencement for life. He recognizes that the hunt for the first living things is the pursuit of a will-o'-the-wisp, and that there never were any first living things either on this earth or anywhere else.

The great lesson that the study of evolution teaches is that Nature knows no beginnings but only change. Evolution is continuous, without beginning and, so far as we can see, without end ; and organic evolution is inseparable from the evolution of the inorganic world.

Astronomers and geologists tell us that the planet on which we live was at one time so hot that the elements of which it is composed existed only in a gaseous condition. As it cooled

it condensed and partially liquefied, and further cooling and
condensation led to the formation of a solid crust, surrounding
a core which is still intensely hot and at any rate partially
liquid, and surrounded by an atmosphere composed of certain
constituents that remain gaseous at comparatively low temperatures. When the temperatures had fallen sufficiently to allow
of the condensation of aqueous vapour in this atmosphere,
rivers began to flow and lakes and oceans to accumulate. It
was then that the existence of living things on the earth's
surface gradually became possible.

All along a process of evolution must have been going on
amongst the constituents of which the cooling earth was
composed. Chemical changes—combinations and decomposi-
tions of all kinds—were constantly taking place in strict
accordance with the physical conditions existing for the time
being, and it was doubtless the chemical reactions between the
more superficially placed constituents of the earth's crust and
the surrounding atmosphere, in the watery medium provided
by the condensation of the aqueous vapour, and under the
influence of the sun's rays, that led, slowly but surely, to the
formation of that curious substance which we call protoplasm
—a substance that occurs in the bodies of all living organisms
and in the absence of which life—as the biologist understands
it—is unknown.

The actual stages that led up to the formation of protoplasm as soon as the conditions at the surface of the cooling
earth became suitable are unknown to us. We are equally
ignorant as to whether the process may have taken place once
or many times. It was, of course, a kind of spontaneous
generation, but a spontaneous generation by slow and gradual
evolution of the inorganic into the organic, and very different
from the sudden conversion of inanimate matter into more or
less highly organized plants and animals that has so often
been imagined. So far as we can yet see, living organisms are
never produced in this way at the present day, but always
arise as the offspring of living parents.

It is the chemical and physical reactions between the protoplasmic organism and its environment that constitutes what we call the " life " of the organism, and the more intense and varied these reactions are the more intense and varied will be the life. I am speaking now of life from the purely mechanistic point of view, but I do not wish to be understood as suggesting that this is the only point of view, or even that it is the most satisfactory point of view, though perhaps it is the only one that a scientific man, as such, can safely adopt, for it is the only one that gives a chance of investigation by purely scientific methods. It is, however, by no means irreconcilable with the views of those who hold that the spiritual aspect of the universe is no less real and no less important than the material. Life, it is true, appears to the pure mechanist to consist entirely of chemical and physical processes, but this may be merely because these are the only processes that are accessible to investigation by the very limited means at his disposal. However it may be with the lower forms of living matter, the case against pure mechanism appears to be conclusive when we come to consider the phenomena of consciousness and thought exhibited by human beings, and the biologist cannot stultify himself by admitting any breach of continuity between the simplest and the most complex of living things.

It would seem, then, that as the chemical evolution of certain carbon compounds progressed they slowly acquired those properties which we regard as distinctive of living beings, though we can by no means say at what particular point they first became alive, for the transition from the not-living to the living condition must have been perfectly gradual. Amongst the properties that are more especially characteristic of protoplasmic organisms is the power of taking in food material from the environment. This not only furnishes fresh supplies of energy but, inasmuch as the food material may be converted by chemical processes into new protoplasm, it leads to growth. Growth, however, is not continuous, but sooner or later results in the multiplication of protoplasmic units, and thus the swelling stream of living matter becomes broken up into individual organisms, each capable of reproducing its own kind, but always liable to further modification, which leads, on the whole, to ever-increasing complexity of structure—though under certain exceptional conditions the evolutionary process may be reversed and result in simplification. In this way we believe that the innumerable host of living things that to-day people the earth in all their endless variety has slowly and gradually come into being, the stream of life constantly branching out in new directions and the individual organisms adapting themselves in structure and function to every situation in which the maintenance of life is possible.

Amongst the more important factors that contribute towards the highly complex process of organic evolution we may safely give precedence to the power of the living proto-plasm to respond to stimuli—that is, to exhibit some change in behaviour as the result of some change in the surrounding conditions, or environment, both being of a more or less definite character. Even comparatively simple organisms, like the Amœba, which look like microscopic specks of semi-transparent jelly, though in reality having a very complex structure, respond quite definitely to such stimuli as contact, light, heat, electricity, and chemical action, contracting themselves into balls or changing the direction of their move-ments as the case may be. Similarly, in the higher plants, a root responds to the stimulus of gravity by growing downwards and a shoot by growing upwards. Not only does the complete organism, or its individual organs, respond in this matter, but the actual building up of its structure, as it develops from the egg, is determined in the same way. The protoplasmic cells divide and their products arrange themselves under the influence of the appropriate stimuli, and if these stimuli be changed beyond narrow limits the change will inevitably be reflected in the structure of the organism. Thus the growth and differentiation of a plant or an animal are merely expressions of its behaviour under the influence of specific stimuli, and the result is to be attributed ultimately to the responsive power of the living protoplasm, a power that seems to be quite capable of explanation in terms of chemical and physical science.

But the living protoplasm is not only capable of responding to stimuli, it is also capable of learning by experience and thus making its responses more and more prompt and accurate each time they are repeated. In other words, it can be educated and can establish habits. According to the experiments of Professor Jennings, this power of learning by experience can be recognized even in the unicellular organisms, and it must, I think, be regarded as a fundamental property of living matter.

Perhaps the most potent stimulus that acts upon the living organism is that of hunger, which leads it to take in food from the external world, and with it fresh supplies of energy, making good what has been dissipated in its various activities and providing the material necessary for growth and reproduction. Hunger is an internal stimulus, arising within the organism itself, and when it is supplemented by the stimuli provided by the presence of suitable food material the organism makes the appropriate response, captures or absorbs the food, and incorporates it in its own body, there to undergo the processes of digestion and assimilation. In ultimate analysis, of course, the act of feeding, even in the higher animals, is performed by all the constituent protoplasmic units, or cells, of which the body is composed, just as it is performed by free-living unicellular organisms such as the Amoeba.

The living organism, then, whether it be a single cell or a complex multicellular plant or animal, learns by repeated experience to obey ever more promptly what we call the feeding instinct. If food be abundant it may be taken in in excess of immediate requirements and stored up in some suitable form for future use. The fats, oils, starches, and so forth, so commonly met with in the cellular tissues of plants and animals, and especially in eggs and seeds, constitute a reserve that can be drawn upon whenever the necessity arises, and this reserve represents so much surplus energy.

It is just this habit of accumulating surplus energy that has made progressive evolution possible. The parent organism accumulates a certain amount of capital which may, at any rate in part, be devoted to giving its offspring a better start in life. The methods by which this result is attained vary endlessly in detail, but the principle is always the same. In the animal kingdom, in the course of evolution, there has been a gradual improvement in the method of transmitting capital from parent to offspring. The first great discovery, so to speak, was the utilization of the egg itself as a storehouse of food material transferred from the body of the parent. This method reaches its climax in the reptiles and birds. The chick embryo in a hen's egg is, indeed, an almost perfect example of a capitalist, and no one would think of calling it " bloated ", for the accumulated capital is all required for perfectly legitimate and indeed necessary purposes.

The mammals, whose ancestors undoubtedly also laid large eggs, made two new departures in this respect, to which we have already had occasion to refer. In the first place they learnt to retain the young within the womb of the mother and feed it there by means of the blood circulation until it reached an advanced stage of development; and, concomitantly with this, they developed milk-glands from the skin, and were thus enabled to continue the feeding process after the young were born. These two methods of transferring capital, or surplus energy, from one generation to the next have proved so successful that they have entirely superseded the older and more primitive method of the birds and reptiles, and the mammalian egg, being no longer used largely as a storehouse for reserve food material, has now reverted to the microscopic size that seems more consistent with its character as a single cell.

It seems sufficiently obvious that the habit of accumulating capital for the benefit of the next generation must lead to progressive evolution, for each successive generation, on the average, gets a slightly better start in life than its predecessor and is thus enabled to carry on its individual development a little further. This conclusion is not open to the objections that have, rightly or wrongly, been urged against Lamarck's theory of the gradual adaptation of the organism to its environment through the inheritance of acquired characters, for whether the supplies of energy be accumulated in the egg-cell in the form of yolk and albumin, or whether they be transferred from parent to offspring by the processes of gestation and lactation, we are able to demonstrate quite clearly the means by which the accumulation is handed on. Unless something occurs to prevent it, it must continue to increase like a sum of money at compound interest, and each generation, in virtue of the inherited experience of its living protoplasm, will tend to accumulate more rapidly than its predecessor.

The mere accumulation of surplus energy cannot, of course, determine particular lines of evolution; it supplies, so to speak, only the motive power that renders progress possible. What it is that directs this progress so as to ensure the close adaptation of the organism to the conditions under which it has to live is another question.

It seems, then, that the accumulation of capital, which is simply another name for surplus energy, is no mere human invention, but an essential condition of progress throughout the organic world. So also is the transmission of capital from parent to child. The great principle involved in capitalism is, perhaps, as old as life itself; it is only the abuse of that principle that has been reserved for mankind.

There is one other primary factor of organic evolution that is at least as old as capitalism, and that is work, for capital and labour have been co-operating ever since the beginning of things, and neither can do without the other. We have already seen that life, from the biological point of view, consists in the constant interchange of energy between the organism and its environment. The organism takes in energy in the form of food and expends it in a great variety of ways in overcoming the antagonism of its surroundings. Now if we define work as the expenditure of energy, which seems to be the only satisfactory definition, it follows immediately, not only that life is dependent on work, but that life and work are in reality one and the same thing, or, to put it more accurately, that life is nothing but work of a particular and very complex kind. The higher the type of organism the more complex is the work in which its life consists and the greater are its capacities for good and evil, pleasure and pain.

It is clear, then, that every living thing must work for its living, for the cessation of work means death. We cannot even attempt to distinguish scientifically between workers and non-workers without realizing at once that the non-workers are all dead and that there can be no need for us to trouble ourselves any further about them.

BOOK II
THE DIVERSITY OF LIFE

THE DIVERSITY OF LIFE

THE first book has emphasized the unity of living things—a unity of structure, of function, of origin and to some extent of history. Now the stress changes from the unity to the differences. The elemental strands on the great distaff which Clotho holds are all fundamentally alike, but on the spindle of Lachesis an almost infinite variety of threads can be spun out of them. It is Evolution which has given rise to all the present variety in the pattern of life, and the problem of how this has happened must be our next quest.

On the surface, the great rule of life seems to be that like produces like. So little difference can be noticed, in the lifetime of a single observer, in the vast majority of species of plants or animals, that it seems only reasonable to assume that no change takes place, or has ever taken place. But natural changes have been detected—in moths, for instance—occurring in the course of a few years, and man has produced many new varieties in plants and animals which are of economic importance ; while in the fossils we have a massive array of evidence to show how great can be the cumulative effect of small variations when these continue through long ages. However, the regularity of the life-pattern is the most striking feature. *How is the continuity of life maintained ?* An answer to this question, and an introduction to our second book, is given by Professor Waddington, who was Chief Geneticist to the National Animal Breeding Research Organization—a reminder of the benefits which applied biology has brought to the farmer and stock-breeder.

CONRAD HAL WADDINGTON

THE BIOLOGICAL ORDER

IF you were a scientist, and sat down to the job of reaching a thorough understanding of living things, what would you consider the most interesting and important questions to study ? The world of animals and plants is bewilderingly complicated. There are a thousand—a million—queer and

interesting things in it. It has taken biologists a long time to
see the wood for the trees, to spot the really basic things
among the mass of detail. In fact, I doubt if we can say that
we have really got there yet, but I will try to tell you something about the direction our ideas are taking nowadays, and
where I guess they may be going in the future.

One line of attack on biology is to follow up the trail
started by the older sciences which deal with non-living things
—the sciences of chemistry and physics. You can ask what
substances go into an animal with its food, what happens to
them when they get inside, what processes do they take part
in, and how do these processes provide the energy the animal
uses to move about, and so on. By studying these things,
you will eventually, if you are clever enough, get some understanding of the calories in your diet, the proteins, carbohydrates, fats and so on that it is made up of, and the vitamins
and hormones and enzymes—all the substances that the body
uses—and you may also find out something about what they
do and how they do it. But, however much you learn in this
way, it will only be a partial account of what a living thing is
like. It will be as though all you knew about a motor car was
about the petrol, oil, and water that go into it, with all the
details you like about the octane rating of the fuel, and anti-
knock dope, the anti-freeze added to the water and so on.
Clearly this would never give you the whole story, unless you
also knew about the cylinders, pistons, carburettor, magneto,
and all the rest of the structure within which the petrol, oil,
and water are put through their paces.

When we are considering living things the situation is,
in some ways, very much the same. The nature of the things
out of which the animal is built, and the processes in which
these substances take part, are only half the story. The other
half must be concerned with the shape and structure of the
machinery on which life depends. This is the problem usually
spoken of as that of " biological order ". Living things are
orderly, arranged in a definite structure, not just a chaotic

conglomeration of fats and sugars and vitamins and so on mixed higgledy-piggledy.

There is nothing very mysterious about the orderliness of an ordinary machine, because we know that the machine has been designed and made by men, who have put it together so that it can do certain definite things. But the orderliness of an animal or plant is obviously something very different. Without going into any questions of who (if anyone) designed it, which would lead us into difficult questions of theology, we can see for ourselves that it behaves quite differently. First it seems to make itself. A living thing starts life as a fertilized egg, a blob of simple-looking jelly, usually a very tiny blob, unless it is attached to a larger mass of foodstuff provided for the young creature, as it is in a hen's egg. This simple-looking egg develops, under its own steam as it were, all the organs of the fully adult animal, arranged in their right places so as to work together as a going concern. How does it do it?

Before trying to answer that question, let us turn to the other peculiar characteristic of biological order. That is its continuity, combined with a slow gradual change over the course of centuries. The ordered structure of one individual perishes and disappears when that individual dies; but it is passed on to reappear in his offspring. It is transmitted by heredity, changing only very slowly in the long process of evolution.

Heredity is a subject which biologists have tackled rather successfully. Anything that can be passed on from parent to offspring must be carried on in one or other of the two single-cells—the egg or the sperm—from which the new individual arises. There is not room for very much in a single cell. Whatever it is which passes on in heredity must be something fairly simple. Actually, it has been shown, in very numerous experiments, that the most important things which are passed on from parent to offspring are a few thousand particles of a special kind which are carried in the cell-nucleus, which is a sort of bag or container for them. These particles are known as genes. I do not mean that the genes are the only things passed on ; the egg contains quite a sizable mass of jelly as well as the genes, although the sperm does not have much besides them to contribute. What I do mean is that the genes are the more important of the two, because if the genes in an animal are abnormal, then the adult which develops will be abnormal, whereas we find very few abnormalities or peculiarities that can be traced back to changes in the rest of the egg. We shall see that the rest of the egg cannot be left out of account: it has quite an important part to play in the whole drama of life, but it is the genes that are mainly responsible for the continuity between one generation and the next.

This property of continuity is perhaps the most important, and also one of the most mysterious, things about them. As the animal grows up, the first two cells form four, and then eight, sixteen and so on, and the genes reproduce themselves exactly, so that each new cell is provided with representatives of each kind of gene. So unless a gene is there in the egg to begin with, it will never appear later, or if one of the original genes in the egg has been altered in some way, and taken an abnormal form, it will go reproducing itself in that abnormal condition. Every gene has the power of making a new gene exactly like itself, ready to hand on to a new cell when one appears. And genes are the only things in cells that we know to have this property—that is what is meant by saying that they are responsible for the continuity of biological order. But besides making another gene like themselves, they have to do something more, they have to produce substances that affect the cell and control the way it develops.

How they do it is quite another question. I have spoken of genes as something rather simple. So they are to the biologist, who is used to thinking of whole animals or plants, or organs such as the liver or kidneys, or even cells such as blood cells or brain cells. But the chemist starts at the other end of the scale. He deals with groups of atoms called molecules—very simple groups, like common salt, or more complicated ones like sugars, made of about half a dozen carbons and oxygens and a dozen or so hydrogens, joined up in a number of different ways. From the chemist's point of view a gene is very complicated. Each gene certainly contains many thousands of atoms, probably even hundreds of thousands. We still do not know at all exactly how big they are, or what shape they have. We still have not been able to see them satisfactorily. They are too small for the best of the old-fashioned microscopes. That is one of the places where the new electron microscope should be able to give us a great deal of help. There are, it is true, many technical difficulties in getting adequate electron pictures of genes, but they are almost certain to be overcome in the next few years, and we may then hope to get a better idea, not only of the over-all size and shape of these fundamental units, but probably also about their internal make-up. A particle with as many atoms in it as a gene has, will very likely prove to be built up out of a number of medium-sized sub-units, and the electron microscope may be able to distinguish them for us.

We are already pretty certain that the gene has two main essential constituents. One of these is a substance known as nucleic acid—that is, the acid of the nucleus—the other is a protein. Which is the more important is anybody's guess. Nucleic acids often seem to turn up wherever a lot of new protein is being formed ; but whether they are just some sort of oil in the works, or whether they determine what kind of protein is to be made, we still do not know. And as for the protein part of the gene—well, proteins are the most characteristic kind of substance in a living body, and they are involved in nearly all living activities. Muscle, for instance, is almost entirely made of protein, and all the more complicated processes of life are directed by substances known as enzymes, which again are proteins. So the continuity of life depends on the way in which the genes, either the nucleic acid part or the protein part, or both, control the formation of new substances in the growing individual.

Let us go back to the question of how the egg produces a complete animal with all its organs arranged in an orderly fashion. This would be quite mysterious if the egg was nothing but a featureless lump of jelly containing a bag full of genes. Actually that is not so. The jelly part of the egg—the part outside the nucleus—is never just a structureless blob. It always has some arrangement into different parts : it has a top and a bottom, a back and a front and so on, which are to some extent different from one another. What seems to happen then is that, as the egg becomes divided into a number of cells, some genes find themselves near the top end, and they react with the material there and cause it to develop into a certain organ, while in another part of the egg the genes react differently because the local materials in that part are different, and so the same genes will cause a different organ to form there. In this way, the whole complicated and orderly structure of the animal is gradually built up, by the reactions between the genes and the original very simple order which we found in the egg in the beginning.

You can see from this why it is that many biologists to-day think that their most important task is to try to understand the way genes react with the other substances surrounding them. It is these reactions which are, in the main, responsible both for the continuity and for the orderliness of living things. But you will also understand that, as the genes are too small to be seen with any ordinary microscope, studying their reactions is no easy matter. But again there is a new device which may come to our help—that is, using the labelled atoms which atomic energy plants are producing for us. How far this will enable us to go on solving the problem, I cannot say. But this question, what kind of reaction genes take part in, and how they control the new substances that are formed, is certainly going to be one of the major preoccupations of science in the next few years at least.

Our next task is that of discovering how Nature, while making sure that each generation is like the one that has gone before, still leaves room for little variations to arise, and to establish themselves if they are favourable, so that in the course of ages new varieties and then new species can arise. *How does heredity work?* This is far too big a question to be answered in one lecture, and pride of place in the " team " reply must go to the Abbé Mendel, although his discoveries, made about the year 1860, were almost completely neglected until 1900.

Here are some extracts from the papers which he read to the Brünn Natural History Society in 1865. The translation is taken from W Bateson's book, *Mendel's Principles of Heredity,* and the extracts illustrate the two primary laws of heredity. The first of these states that heredity works by means of indivisible determinants, or " genes " as they are now called. Formerly it was thought that there was a blending or dilution of characters. If, for example, a white man married a black woman their children would have a skin-colour midway between the two, and if one of these married a white (or black) spouse, their children again would show a blending of their parents' colour. This did prove to be the case fairly often, but many surprising anomalies occurred. Thus the marriage of two mulattoes might produce children of any shade between white and black. As Morgan worded Mendel's first law, " The units contributed by two parents separate in the germ cells of their offspring without having had any influence on each other ". The second law stresses the independence of these genes. With certain exceptions (like the sex-linked characters of the poultry farmer) each gene plays an independent part in the machinery of inheritance.

GREGOR MENDEL

EXPERIMENTS IN PLANT HYBRIDIZATION

EXPERIENCE of artificial fertilization, such as is effected with ornamental plants in order to obtain new variations in colour, had led to the experiments which will here be discussed. The striking regularity with which the same hybrid forms always reappeared whenever fertilization took place between the same species induced further experiments to be undertaken, the object of which was to follow up the developments of the hybrids in their progeny. . . .

That, so far, no generally applicable law governing the formation and development of hybrids has been successfully formulated can hardly be wondered at by anyone who is acquainted with the extent of the task, and can appreciate the difficulties with which experiments of this class have to contend. A final decision can only be arrived at when we shall have before us the results of detailed experiments made on plants belonging to the most diverse orders.

Those who survey the work done in this department will arrive at the conviction that, among all the numerous experiments made, not one has been carried out to such an extent and in such a way as to make it possible to determine the number of different forms under which the offspring of hybrids appear, or to arrange these forms with certainty according to their separate generations, or definitely to ascertain their statistical relations. It requires indeed some courage to undertake a labour of such far-reaching extent ; this appears, however, to be the only right way by which we can finally reach the solution of a question the importance of which cannot be over-estimated in connexion with the history of the evolution of organic forms.

The paper now presented records the results of such a detailed experiment. This experiment was practically confined to a small plant group, and is now, after eight years' pursuit, concluded in all essentials. Whether the plan upon which the separate experiments were conducted and carried out was the best suited to attain the desired end is left to the friendly decision of the reader.

[Mendel went on to explain why he chose the pea family for his experiments (a fortunate choice, by the way) and to list the characters which he selected for his crossing experiments. These latter were: (1) the shape of the ripe seeds, round or wrinkled ; (2) the colour of these seeds, yellow, orange, or green ; (3) the colour of the seed-coat, white, grey, or brown ; (4) the shape of the ripe pods ; (5) the colour of the unripe pods ; (6) the position of the flowers on the stem ; and (7) the length of the stem. In all his crossings care was taken to make them both ways, the pollen playing the part in one set of experiments which the seed played in another set, and vice versa.]

THE FORMS OF THE HYBRIDS

Experiments which in previous years were made with ornamental plants have already afforded evidence that the hybrids, as a rule, are not exactly intermediate between the parental species. . . . This is . . . . the case with the pea-hybrids. In the case of each of the seven crosses the hybrid-character resembles that of one of the parental forms so closely that the other either escapes observation completely or cannot be detected with certainty. This circumstance is of great importance in the determination and classification of the forms under which the offspring of the hybrids appear. Henceforth in this paper those characters which are transmitted entire, or almost unchanged in the hybridization, and therefore in themselves constitute the characters of the hybrid, are termed the *dominant*, and those which become latent in the process *recessive*. The expression " recessive " has been chosen because the characters thereby designated withdraw or entirely disappear in the hybrids, but nevertheless reappear unchanged in their progeny, as will be demonstrated later on.

It was furthermore shown by the whole of the experiments that it is perfectly immaterial whether the dominant character belong to the seed-bearer or to the pollen-parent ; the form of the hybrid remains identical in both cases. . . .

Of the differentiating characters which were used in the experiments the following are dominant :

(1) The round or roundish form of the seed with or without shallow depressions.

(2) The yellow colour of the seed albumen.
(3) The grey, grey-brown, or leather-brown colour of the seed-coat, in association with violet-red blossoms and reddish spots in the leaf axils.
(4) The simply inflated form of the pod.
(5) The green colouring of the unripe pod in association with the same colour in the stems, the leaf-veins, and the calyx.
(6) The distribution of the flowers along the stem.
(7) The greater length of stem. . . .

THE FIRST GENERATION (BRED) FROM THE HYBRIDS

In this generation there reappear, together with the dominant characters, also the recessive ones with their peculiarities fully developed, and this occurs in the definitely expressed average proportion of three to one, so that among four plants of this generation three display the dominant character and one the recessive. This relates without exception to all the characters which were investigated in the experiments. . . . *Transitional forms were not observed in any experiment.* . . .

. . . The relative numbers which were obtained for each pair of differentiating characters are as follows :

EXPT. 1. *Form of Seed.*—From 253 hybrids 7324 seeds were obtained in the second trial year. Among them were 5474 round or roundish ones and 1850 angular wrinkled ones. Therefrom the ratio 2·96 to 1 is deduced.

EXPT. 2. *Colour of Albumen.*—258 plants yielded 8023 seeds, 6022 yellow, and 2001 green ; their ratio, therefore, is as 3·01 to 1. . . .

EXPT. 3. *Colour of the Seed-coats.*—Among 929 plants 705 bore violet-red flowers and grey-brown seed-coats, giving the proportion 3·15 to 1.

EXPT. 4. *Form of Pods.*—Of 1181 plants 882 had them simply inflated, and in 299 they were constricted. Resulting ratio, 2·95 to 1.

EXPT. 5. *Colour of the Unripe Pods.*—The number of trial plants was 580, of which 428 had green pods and 152 yellow ones. Consequently these stand in the ratio 2·82 to 1.

EXPT. 6. *Position of Flowers.*—Among 858 cases 651 had inflorescences axial and 207 terminal. Ratio, 3·14 to 1.

EXPT. 7. *Length of Stem.*—Out of 1064 plants, in 787 cases the stem was long, and in 277 short. Hence a mutual ratio of 2·84 to 1. In this experiment the dwarfed plants were carefully lifted and transferred to a special bed. This precaution was necessary, as otherwise they would have perished through being overgrown by their tall relatives. Even in their quite young state they can be easily picked out by their compact growth and thick dark-green foliage.

If now the results of the whole of the experiments be brought together, there is found, as between the number of forms with the dominant and recessive characters, an average ratio of 2·98 to 1, or 3 to 1.

The dominant character can here have a *double signification* —viz. that of a parental character, or a hybrid character. In which of the two significations it appears in each separate case can only be determined by the following generation. As a parental character it must pass over unchanged to the whole of the offspring; as a hybrid-character, on the other hand, it must maintain the same behaviour as in the first generation.

THE SECOND GENERATION (BRED) FROM THE HYBRIDS

Those forms which in the first generation exhibit the recessive character do not further vary in the second generation as regards this character; they remain constant in their offspring.

It is otherwise with those which possess the dominant character in the first generation (bred from the hybrids). Of these *two*-thirds yield offspring which display the dominant and recessive characters in the proportion of 3 to 1, and thereby show exactly the same ratio as the hybrid forms, while only *one*-third remains with the dominant character constant.

THE OFFSPRING OF HYBRIDS IN WHICH SEVERAL
DIFFERENTIATING CHARACTERS ARE ASSOCIATED

In the experiments above described, plants were used which differed only in one essential character. The next task consisted in ascertaining whether the law of development discovered in these applied to each pair of differentiating characters when several diverse characters are united in the hybrid by crossing. As regards the form of the hybrids in these cases, the experiments showed throughout that this invariably more nearly approaches to that one of the two parental plants which possesses the greater number of dominant characters. . . . Should one of the two parental types possess only dominant characters, then the hybrid is scarcely or not at all distinguishable from it.

Two experiments were made with a considerable number of plants. In the first experiment the parental plants differed in the form of the seed and in the colour of the albumen ; in the second in the form of the seed, in the colour of the albumen, and in the colour of the seed-coats. Experiments with seed-characters give the result in the simplest and most certain way.

In addition, further experiments were made with a smaller number of experimental plants in which the remaining characters by twos and threes were united as hybrids ; all yielded approximately the same results. There is therefore no doubt that for the whole of the characters involved in the experiments the principle applies that *the offspring of the hybrids in which several essentially different characters are combined exhibit the terms of a series of combinations, in which the developmental series for each pair of differentiating characters are united.* It is demonstrated at the same time that *the relation of each pair of different characters in hybrid union is independent of the other differences in the two original parental stocks. . . .*

All constant combinations which in peas are possible by the combination of the said seven differentiating characters were actually obtained by repeated crossing . . . . Thereby is . . . given the practical proof *that the constant characters which appear in the several varieties of a group of plants may be obtained in all the associations which are possible according to the (mathematical) laws of combination, by means of repeated artificial fertilization. . . .*

If we endeavour to collate in a brief form the results arrived at, we find that those differentiating characters, which admit of easy and certain recognition in the experimental plants, all behave exactly alike in their hybrid associations. The offspring of the hybrids of each pair of differentiating characters are, one-half, hybrid again, while the other half are constant in equal proportions, having the characters of the seed and pollen parents respectively. If several differentiating characters are combined by cross-fertilization in a hybrid, the resulting offspring form the terms of a combination series in which the combination series for each pair of differentiating characters are united.

The uniformity of behaviour shown by the whole of the characters submitted to experiment permits, and fully justifies, the acceptance of the principle that a similar relation exists in the other characters which appear less sharply defined in plants, and therefore could not be included in the separate experiments.

Mendel, as we have said, was fortunate in his choice of the garden pea for his experiments, mainly because of the clear-cut dominance of the yellow colour and the tall habit in these plants. He was thus able to formulate the basic principles of heredity, unhampered by the complications which other plants or animals might have thrown in his way. Out of those early experiments a whole new science of *Genetics* has developed. In the early years of the twentieth century his work was repeated and extended by biologists in a large number of countries, and it was shown that his laws applied to a very wide range both of plants and animals. Biologists are rightly proud of the progress that has been made, and for the second part of the answer to our question we go to Dr. Julian Huxley. Here are some extracts from his book, *Soviet Genetics and World Science.*

Within recent years, biology in Soviet Russia has had to fit itself into the framework of Marxist theory. The idea of competition between individuals of the same species, for example, could not be allowed, and that ruled out any explanation of evolution working through selection, as suggested by Darwin. So a different machinery for evolution had to be "discovered". Most of it was invented by the plant-breeder, Lysenko, who adopted and modified Lamarck's theory of the inheritance of acquired characters. A battle-royal developed between Russian and Western geneticists. We do not propose to review the battle here. Science must depend on observations and experiments which can be verified by any sufficiently skilled worker, and verification of Lysenko's theories has been singularly lacking. However, the need to defend the Western views led to Dr. Huxley stating the basic principles of genetics with admirable clarity.

SIR JULIAN HUXLEY

GENETICS AS A SCIENCE

CHROMOSOMES were first properly investigated in the 1880s. Before 1890 the basic facts concerning their behaviour in cell-fusion and in sexual reproduction had been established, and by 1900 they had been demonstrated in the nuclei of the cells of all higher animals and plants examined for them.

Each species was found to have a constant number of chromosomes, normally in pairs. Thus the number of chromosome-pairs in man is 24; in the famous fruit-fly, *Drosophila melanogaster*, of which many millions have now been bred and counted in genetical experiments, it is 4; and 7 in the garden pea used by Mendel.

We may conveniently compare the chromosomes to playing cards. Then, in all higher organisms, the fertilized egg and all the cells of the individual to which it gives rise, typically have two packs or complete sets of chromosomes in their nuclei. But before the gametes [the marrying cells of ova and sperm] are formed, a complicated process of pairing and separation takes place, so that each gamete has only one pack of chromosomes, but a complete pack containing one of every kind. This process is called *meiosis*, meaning reduction (to half the number of chromosomes). And of course fertilization brings two packs together, one from the egg and the other from the sperm, and restores the double number once more.

It was soon realized that Mendel's two laws would be realized if the factors of heredity (or *genes*, as they were later conveniently called) were lodged in the chromosomes, and if each kind of gene could exist in a number of slightly different forms, or *allels*. The first law applies to one pair of genes (to be quite precise, to the pair of allels of one kind of gene) in one kind of chromosome, the second law to two or more pairs of genes lodged in different kinds of chromosomes.

Morgan and his school then suggested that, if the genes were strung out in a single row within the chromosomes, the laws of linkage should apply to two or more pairs of genes which were lodged in one and the same kind of chromosome. And more detailed observation of the behaviour of chromosomes later showed that at meiosis the members of each chromosome pair twine round each other and exchange segments of their length, in precisely the way which would account for the numerical facts of linkage.

As a result, the chromosome theory of heredity was born, combining the facts drawn from breeding experiments with those derived from microscopic observation. It became possible to construct " chromosome maps " giving the relative positions of the genes (factors) within the chromosomes. This was done by taking the closeness of linkage between different kinds of genes as some measure of the physical distance between them along the chromosome's length.

The chromosome theory, like any scientific theory, has predictive functions : if it is true, then certain results should follow. These predictions can be tested. When fulfilled, they provide further confirmation for the general framework of concepts which we call the theory : when they are not fulfilled exactly, the theory has to be amplified or slightly modified. Of course, if results are obtained which are incompatible with the theory, the theory has to be scrapped and a new one substituted (as Newtonian mechanics has been replaced by relativity theory). But so far nothing of the kind has happened with the chromosome theory ; on the contrary, it is now buttressed with so many new supporting and confirmatory facts that it would appear impregnable. . . .

What are perhaps the most remarkable predictions only became possible later, with the discovery of the salivary gland chromosomes of Drosophila, the fruit-fly. Fruit-flies are ideal for genetic experiments, but their ordinary chromosomes are so unusually small as to be very unfavourable for microscopic study. Then one day it was discovered that in the salivary glands the cells contained giant chromosomes, so enormous that their fine structure can be seen under the microscope. This fine structure consists of a series of discs and bands, and each chromosome has a particular arrangement of these. Thus, in addition to the theoretical maps of the genes deduced from linkage experiments, it was now possible to make actual maps of the bands and discs derived from observation. What is more, by brilliant but laborious experiments, it proved possible to equate the two, so that now in any textbook of genetics you will find maps which give not only the actual structure of the fruit-fly's chromosomes, but also the position of hundreds of genes known from breeding experiments. (In a number of cases, indeed, it has been possible to define the precise limits of individual genes on the map, as well as their general position.)

Now breeding experiments had occasionally given results which did not tally with the normal gene-map. Quite early, the Morgan school had put forward the hypothesis that these aberrant results were due to small sectors of a chromosome becoming misplaced—either reversed end to end, or detached from their normal position to become attached to other chromosomes. All the facts proved capable of being interpreted on this basis, but there were critics who felt that the suggested explanation was a bit far-fetched. However, with the discovery of the salivary gland chromosomes, it became possible to test the hypothesis, and it was a triumph for the chromosome theory when observation confirmed prophecy. When, for instance, the prophecy was that a piece of chromosome had been translocated elsewhere, lo and behold, that part of the salivary gland structure *was* found in the new place.

An even more spectacular confirmation happens with inversions (pieces of chromosome reversed end to end). It was previously known that at meiosis the members of a chromosome pair joined up with great exactitude, each gene pairing up with its opposite number. Now the only way that this could happen when a section of one chromosome is reversed would be for the section to form a twisted loop, fitting against an untwisted bulge of the normal chromosome. And this conformation is then visible through the microscope (and, I may add, never otherwise found).

Meanwhile a vast amount of patient work had served to establish two exceedingly important facts about Mendelism, using Mendelism in the extended sense to mean " particulate chromosome inheritance "—inheritance by means of the transmission of definite particles or units of living substance, which are lodged in chromosomes, which maintain their identity from generation to generation, which segregate, and which can be recombined in various ways. (It now seems that a few properties are transmitted by so-called *plasmagenes*—units in the general protoplasm instead of in the chromosomes, which accordingly do not segregate and may be present in varying numbers in each cell ; but the great bulk of inheritance in the great majority of organisms is both particulate and chromosomal.)

The first fact was that Mendelism was not an exceptional phenomenon, but universal, at any rate in all but some of the lowest and simplest microscopic organisms. And the second was that it applies not only to obvious differences such as the difference between tall and dwarf peas or albino and coloured rabbits, but to all kinds of character-differences, whether large or small.

The second fact is the smallness of the differences which may mendelize ; this is very important. Thus one gene may act as a modifier of another, producing a slightly greater or smaller size, for instance, or a slightly darker or slightly lighter colour ; or two or more genes (" multiple genes ") may combine more or less equally in the production of some effect : and one character may be affected by a large number of these modifying and multiple genes.

The various genes can usually be disentangled by suitable (though often laborious) experiment. But the net effect they produce in nature is often one of continuous variation, such as a complete range of size from small to large, instead of discontinuous differences, such as that between Mendel's dwarf and tall peas.

One important point about particulate inheritance is that whenever a cross is made between two genetically different strains, and the F_1 (first generation hybrid) offspring is capable of normal sexual reproduction, the F_2 (second generation) will show a greater range of variation than the F_1. We already saw this with the cross between green and yellow peas, but the principle is of universal application, and often (as with the cross between " yellow tall " and " green dwarf " peas) produces new types not seen in either parent or in the F_1.

This greater variation of F_2 would be impossible on any theory of blending inheritance, though it can be predicted as an inevitable result of particulate chromosomal inheritance.

Sometimes so many pairs of genes are involved in a cross that it is impossible to disentangle them individually. But even then, if we find greater variation in F_2 than F_1, this is evidence of particulate chromosomal inheritance—*i.e.* of the differences between the two parents being due to genes in the chromosomes.

Many people have been puzzled by the wide differences, obviously of genetic nature, which can exist between the children of the same parents. Every one will know families where brothers or sisters differ strikingly from each other, as well as from either parent, in colouring, temperament, stature, intellectual ability, features, or body-build, and where environmental influence can be ruled out as insufficient to produce such large differences. However, such facts are the natural and inevitable consequences of Mendelian segregation and recombination, while they cannot be accounted for on the basis of any other theory of heredity.

The full significance of these facts can only be grasped after we have considered another set of facts recently established—the facts concerning mutation. Mutation is a term which has been given many meanings. However, it has now received a precise and scientific connotation, thanks to the work of neo-Mendelian genetics. It may perhaps be best defined, or rather best described as the result of inexactitude in the process of self-reproduction by the material basis of heredity, the genes and chromosomes.

The genes possess the essential property of life, namely self-reproduction, or preferably self-copying (indeed, if we include plasmagenes, it is probable that they are the only elements in the organism which possess this property). At a certain stage in cell-growth, the string of genes becomes two strings, and one passes into each of the two daughter-cells. Every now and then, however, the copy is not quite exact, and then the new gene produces a different effect in development —absence versus presence of pigment, a different colour, a different hair-form, and so forth. It is by these new effects that mutations can be detected. Then breeding experiments will show whether the effect is due to a mutation in a gene, and not only that, but what particular gene has mutated. Even though a gene has mutated, it still retains the power of self-copying, so that the mutation is self-perpetuating—apart of course from further mutations which may affect it, either changing it back into its original form (" reverse mutation ") or into something new.

This inexact copying of genes is called gene mutation. When large numbers of organisms are bred under controlled conditions, it is possible to tell when a new mutation occurs ; and this has been done with fruit flies, maize, and a few other organisms. It turns out that, for any particular gene, mutation is a rare phenomenon, though its frequency is different for different genes, as well as varying with conditions such as temperature ; that it may produce any change from death or gross disability, through marked effects like albinism or dwarfism, down to trifling quantitative differences ; that even when the differences are trifling, they are always definite in extent—measurable steps of change ; that it is recurrent, in that the same kind of gene is subject to the same kinds of mutation, time after time ; that it is intensely localized, since although two genes of the same kind are always present in all normal cells, mutation never affects both, but only one ; and that it is a random process, in that mutations taken as a whole bear no functional or other relation to the needs of the organism or to the environment in which it finds itself.

All the mutations first studied were spontaneous—they just occurred, and occurred with equal frequency whether the cultures were subjected to special treatment or not. In 1927, however, H. J. Muller made the epoch-making discovery that mutations could be artificially produced by X-rays, and at a rate many times higher than spontaneous mutation. Later experiments (all again involving numbers of facts) have shown that many kinds of radiations are effective in producing mutations, and also certain chemical agents such as mustard gas. In general, artificial mutations are similar to (and often identical with) spontaneous mutations, both in the effects they produce and in the fact that they are random.

I must also mention chromosome-mutations. These differ from gene-mutations in that they do not represent a change in the nature (presumably the chemical structure) of the gene, but quantitative additions or subtractions to or from the entire set of genes which make up the hereditary constitution. One kind of chromosome-mutation we have already referred to—the addition of entire sets of chromosomes and their contained genes. The most familiar is tetraploidy, the doubling of the whole outfit, providing four instead of two representatives of each kind of chromosome (and therefore of each kind of gene-unit). We also find the addition or subtraction of single whole chromosomes or of small bits of chromosomes ; and all these changes produce some effects on the characters of the organism.

Gene-mutations, however, are the most important, for they provide the bulk of the building-blocks of evolutionary change. Once a gene has mutated, it goes on with its self-copying, so that a mutation can persist in the stock for a longer or shorter time. If the effects it produces are harmful, it will eventually be eliminated by natural selection ; if they are helpful the possessors of the gene will be favoured by natural selection, so that the mutant gene will become more and more abundant in the species, and will eventually oust its " normal " parent, itself becoming normal for the stock. Recessive genes of many kinds, but each kind represented only in small numbers, are always to be found in any species of animal and plant (apart from those which practise self-fertilization or close inbreeding, or have abandoned sexual reproduction for some asexual method). These must be the products of past mutation. Their frequency depends on the precise balance between their rate of elimination by natural selection on the one hand, and on the other, their " mutation pressure ", or the rate at which they are produced by new mutations—i.e. new failures of self-copying—in the " normal " genes of which they are the partners. If conditions change, they may become advantageous to the species and increase in number until they become " normal "; or in some cases, two or more mutant genes, each by itself neutral or slightly deleterious, may in combination prove favourable. The stock of mutant genes carried in the chromosomes constitutes the evolutionary reserve of the organism.

Another very important fact is the following : that in general it is the small mutations, producing only slight effects, which are important in the life of the species and valuable for evolution.

On the basis of these facts, a general neo-Mendelian theory could now be framed, including the following main points.

(1) Almost all heritable differences between individuals " mendelize "—that is to say, their distribution in inheritance depends on the behaviour of the chromosomes, which contain the genes, which in their turn are the material units of heredity.

(2) All changes in the hereditary constitution are mutations—*i.e.* changes of definite extent ; these changes are usually within individual genes, but sometimes involve the addition or subtraction of whole genes or strings of genes, up to chromosomes and entire chromosome-sets. Thus all mendelizing differences between individuals—in other words, almost all the heritable variations of plants and animals—owe their origin to past mutation.

(3) Evolutionary change, whether large or small, is almost entirely due to the differential survival of mutant genes and their combinations, harmful ones being eliminated and useful ones being favoured and eventually becoming normal for the stock, through the automatic operation of natural selection. The material basis of evolution is thus the differential survival of genes and gene-complexes, and mutations are the quanta of evolutionary change.

(4) Mutations and the mendelizing effects they produce, though always definite in extent, may be very small, and these small mutations are the more important in evolution. As a result, the variation actually found in a species of animal or plant is generally continuous, without sharp breaks, even though it is based on discontinuous differences in the genes. (In the same sort of way, matter and energy appear continuous to us, although they are actually composed of discrete units—molecules, atoms, and subatomic particles, together with energy-quanta; *e.g.* the " uniform " pressure of a gas is the resultant of the very variable individual behaviour of all its myriad constituent unit-molecules.)

The earlier mendelian geneticists, employing large mutations in their breeding experiments, had somewhat naïvely supposed that nature did the same in its gigantic experiment which we call evolution. They supposed that evolution proceeded by discontinuous jumps. It was natural for them to use sharply contrasting differences for this phase of their work, for these were much easier material with which to establish the mendelian laws and their general applicability. But they made the mistake of thinking that such differences were as important for their animal and plant possessors as for the human experimenters.

For meanwhile the facts of nature spoke in the opposite sense. For one thing, when large samples of animals or plants were measured, it was found that their variation, whether in size or shape or colour, was usually continuous, without trace of breaks, or gaps. And for another, the fossil remains which were unearthed and investigated by palaeontologists, showed that the actual course of evolution had been gradual too—no sudden transformations or leaps into a new type but a slow and continuous process.

The development of a human being out of a microscopic egg-cell was so wonderful to Dendy as to be a " kind of miracle ". We have learnt much about this miracle in recent years, but the wonder still remains. Before we go any further with our study of evolution, let us look more closely at this process of development from egg to adult. Our guide will be Michael Abercrombie, Lecturer in Embryology at University College, London, and in his essay we shall find another example of the unity of life, for the processes he describes are " to a surprising extent " common to all animals. We shall also see *why development from a fertilized egg is the most efficient method of reproduction* —an important factor in the story of evolution. Formerly it was thought that every embryo recapitulated the evolutionary history of its ancestors, but now it is clear that, in Abercrombie's words, " it merely provides evidence for reconstructing its ancestral history, not, as the recapitulation theory implied, the history itself all written out "

MICHAEL ABERCROMBIE

FROM EGG TO ADULT

EVERY adult animal, with few exceptions, has developed from an egg. Adult and egg could hardly be more different from each other. An egg is a single cell, a fraction of the adult size, almost devoid of visible structure. An adult consists of numerous cells (we are leaving Protozoa out of this essay) as well as much material interspersed between the cells ; and it is visibly immensely complicated, always with a multiplicity of different structures woven into an intricate pattern. The egg is motionless, the adult has an elaborate system of behaviour. The change from egg to adult requires therefore an extreme transformation : a transformation so extraordinary that many early embryologists refused to face its possibility. They preferred to imagine development as a mere enlargement of a minute replica of the adult structure, located, according to fancy, in sperm or in egg.

Now that we have adequate microscopes we must accept the existence of the transformation and attempt to analyse it. But before we do so we must ask (sometimes a little peevishly) why development is so drastic a remodelling. What is gained by starting from a single cell? Admitting the necessity for reproduction, why should not an animal simply release an already highly organized fragment of itself? Such an offspring could then develop into a new adult without the necessity of building from the very foundations. Reproduction in this *vegetative* way does indeed occur in some groups of animals, and more prominently in plants. But from its relative infrequency we may infer that it is at a disadvantage compared with the sexual method, in which the starting-point of development, the fertilized egg, results from fusion of a sperm with an unfertilized egg.

Consideration of this problem means that we start embryology, like so much else in biology, by discussing evolution. The advantage of the sexual method lies in the fact that the act of fertilization brings together heritable material (the genes contained in the nuclei) from two individuals, the two parents, and combines it in one offspring. The result is that favourable genes which have originated (by mutation, *i.e.* a sudden and rare change in the properties of a pre-existing gene) in separate individuals can become associated in one individual. And new combinations of those genes whose value lies in their interaction, genes which by themselves are of little advantage, can similarly be realized. Where, as in vegetative reproduction, the making of new organisms is an entirely one-animal show, combination of genes in this way is impossible. Favourable new genes can indeed arise by mutation, but if, as is probable, they arise in separate individuals, they can only compete with each other, which is extremely wasteful. They can be brought together in one and the same individual only if all of them originate by mutation in one individual or its progeny : at the best a slow business. Sexual reproduction thus confers a gain in the flexibility and efficiency of evolutionary advance which has apparently greatly outweighed the disadvantages of its complexity, except perhaps for occasional well-adapted species in conditions of favourable stability. But why not, perhaps, have fertilization, but of an already highly organized multicellular body ? It would certainly need a complicated mechanism to ensure such fertilization ; and a complicated mechanism to prevent the organism becoming a complete hodge-podge of genetic constitutions (a grave disadvantage to effective evolution). A simplification of embryonic development which reproduction of this kind might involve has evidently not counterbalanced its disadvantages. Such a process has, in fact, never been established amongst animals. Probably sexual reproduction by fusion of two single cells, since it is common amongst Protozoa, antedates the evolution of multicellular organization, and animals have stuck to this original method. More and more complicated adult forms have then evolved by a change in processes of embryonic development, bridging the widening gap between egg and adult. The act of fertilization has become not only a method of recombining genes, but a stimulus which sets off an elaborate developmental mechanism.

Let us now consider in outline the visible processes by which animals solve the problem of getting from fertilized egg to adulthood. To a surprising extent these processes are common to all animals, so that I can describe them in very general terms. But it should be borne in mind that living things are extraordinary diverse and that there are likely to be exceptions to most of my generalizations.

ACTIVATION

We have seen that embryonic development is necessitated by the evolutionary importance of the fusion at fertilization of two nuclei from different parents. But although development is *normally* set going by fertilization, it does not necessarily depend on it. The cytoplasm of the ripe unfertilized egg contains for a short time an organization which can be activated in a variety of different ways, only one of which is fertilization by a sperm. Once activated, it can entail all the subsequent stages of development. Unfertilized rabbit eggs, for instance, have been activated by cooling, and development has proceeded to the birth of normal young. Innumerable frog and sea-urchin eggs have been set developing without benefit of sperm. The only difficulty about rearing such fatherless or *parthenogenetic* embryos, as they are called, is that, lacking the sperm nucleus (which like that of the un-fertilized egg is really only a half nucleus), they are short of nuclear material. Consequently, unless the egg nucleus succeeds in doubling itself, which in a proportion of cases it does, the embryo usually dies.

Activation of eggs in the absence of sperm is of course also exemplified in the natural parthenogenesis of aphids and other animals. Here the advantages of making radically new combinations of genes are sacrificed just as in vegetative reproduction; though in nearly all species only temporarily, during favourable conditions. The gain is speeding-up output.

At one time it seemed that a lot could be learned about the organization of the unfertilized and fertilized egg by a study of how artificial activation was evoked. Unfortunately an extraordinary diversity of agents proved capable of starting off development. In some instances several different agents are all effective on members of a single species. Heat, cold, acid, alkali, pricking with a needle, are only a few of the effective treatments. We are embarrassed by riches, and nothing can therefore yet be confidently inferred about what the process of activation is. . . .

We do not know what the organization of any egg is, although there are some visible indications and many experimental demonstrations of its presence. But certainly there are considerable differences between eggs of different species, though extended investigation will probably show that, at least before fertilization, there is also an underlying similarity of organization.

CLEAVAGE

Within a day or two, at the most, the activated egg has embarked on the first stage of its development—one common to all animals. It is a brief phase in which the massive single initial cell is fragmented into smaller cells, much nearer the size of normal body cells; and little else visible happens. The usual process of cleavage is as follows. The egg nucleus divides by mitosis; and the cytoplasm is correspondingly constricted into two by an in-growing groove of the surface layer. The two cells resulting from division, after a rest of a few hours, divide again; and so the process continues. If the egg has a region very rich in yolk, destined purely for embryonic food, then this part may not undergo cleavage. In a bird's egg, for instance, only a tiny cap of protoplasm divides up, the bulk of the yolk remaining unchanged.

The cleavage divisions frequently have a striking temporal and spatial pattern. The cell-divisions and the intervening periods of rest often, to begin with, proceed in a rhythm simultaneously affecting all the cells; so that periodically the cell-number suddenly doubles, then remains constant for a time, then doubles again. The sizes of the two cells produced by any division may be very different; and the plane in which the constricting cytoplasmic groove forms, and consequently the position of the daughter-cells, may vary in different divisions. By orderly arrangement of these variables during cleavage a very precise pattern of cells, of certain sizes in certain positions, may be produced in every individual of a species at a corresponding stage of development. Such characteristic patterns are particularly found among the molluscs and segmented worms (Annelida). In these groups, not only is the cleavage pattern of individuals of one species the same, but there is a fundamental similarity

of pattern characterizing most species. The same fundamental pattern (called *spiral cleavage* because of the oblique
direction of many of the divisions ; the actual pattern of
cells resulting is not a spiral) is found also in other groups,
such as the flatworms (Platyhelminthes). This is powerful
evidence of the relationship of these groups. In many other
groups, however, including the vertebrates, such constant
and fundamental patterns are absent.

The cleavage cells usually adhere together by a sticky
substance which covers their surfaces, or forms a coat round
the whole outer surface of the embryo. It is rather remarkable
that in some animals, including marsupials, the new cells
temporarily fall apart and may for a time be widely separated :
though they are kept from dispersing altogether by the
outer protective membranes which we noted around the egg.
But though the new cells usually adhere in places, it is only
rarely that they remain tightly packed together in a solid
mass. Usually interstices appear internally between the cells,
though the superficial cells remain as a continuous covering.
The interstices run together in many animals and form an
internal cavity. This may swell to a considerable size, and
the cells may eventually form merely a thin membrane, one
cell thick, round a bloated fluid space. The fluid itself is
probably partly a secretion of the cells, more or less mixed
with fluid from the embryo's surroundings, which may be,
for instance, sea-water or, where the embryo develops within
the mother, maternal body-fluid.

As a result of cleavage, therefore, we now have a cluster of
cells, between fifty and a few thousand according to species,
usually arranged as a hollow ball. There has been no particular
change of external shape, no particular increase in size which
is not accounted for by the swelling of the internal cavity, no
very striking advance in structure towards the adult form.
What are we to make of this almost universally occurring
cleavage process ? As a tentative suggestion, we may keep in
mind the possibility that its significance lies in the rapid

increase of cell surfaces that comes from increasing cell numbers. Interaction between cell surfaces is perhaps the key to a good deal of subsequent development.

It is interesting in this connexion that in a few segmented worms it has been possible to suppress cleavage experimentally, while permitting a little development to proceed. Development is then mainly confined to the external layer, where cilia and other structures appear distinctly reminiscent of a normal embryo. Internal organs are absent.

Gastrulation

Just as the phase of cleavage is one of universal cell-division, so the next brief phase, for rather remote historical reasons called gastrulation, is one of universal cell-movement. Masses of the cells resulting from cleavage now stream hither and thither in orderly migrations. Cell-division does not, however, cease during this period, but it becomes as a rule sporadic. The courses of the mass movements have been mapped in a number of animals, chiefly vertebrates, by staining parts of embryos with minute patches of harmless dye. These marks are put on at the end of the cleavage period, and they travel about during the cell-migrations, indicating the routes taken by the cells which bear them. Before such methods had been invented, embryologists were forced to disguise their ignorance by invoking especially active, but none the less imaginary, local growth by cell-division to account for the striking changes in form. The concentration of cell-divisions they sometimes described in certain regions of the embryo cannot be found by eyes whose minds have no such pre-conceptions. The moral which embryologists are learning from this curious episode is that if you want to know how many there are of anything, count them.

The main feature of the orderly cell-migrations is that material which at the end of cleavage was at the surface is carried inside the embryo. There it will eventually form most of the internal organs, particularly the entire gut, but many others too. This is an altogether surprising way of acquiring internal organs. It would seem a simpler method to break up the material of the egg during cleavage in such a way that the internal cytoplasm remains in cells internally placed, and there becomes transformed into internal organs ; while the external cytoplasm is appropriately placed to form the skin. But even when, instead of a single layer of cells, we get a thick wall with internal and external cells surrounding the inner cavity (and that is characteristic of many vertebrates) the obvious very rarely happens ; the material which will make the internal organs still to a large extent streams in from the surface. But there are exceptions. In some members of a very primitive group, the Cœlenterates (jelly-fish, anemones, corals, etc.) which differ from all other multicellular animals except the sponges in having no real organs (no heart or massive muscles or kidneys, for instance) between gut and skin, the material for the gut originates in the simplest way, by differentiation of cytoplasm which has been internal throughout cleavage.

The cells which move from surface to interior during gastrulation do so mainly in one restricted region, not sporadically (again with the exception of a few Cœlenterates). This is equivalent to saying that the cell-migrations are mass migrations. A whole continuous region of the surface disappears inside. It may do so by a dimpling in of the wall of the internal cavity which is so commonly present at the end of cleavage. Or the cells may creep in as a solid mass without much folding of the surface. Or there is a mixture of the two. The place of the immigrant cells is usually taken by an expansion of what will be the future outermost layer of the animal, so that the external size and shape of the embryo does not greatly change. And this expansion of the outer layer may in some species be a major factor in enclosing the future internal material, especially where this

is very massive owing to quantities of yolk. There is great
variety in the course of these movements, and of contribu-
tory rearrangements of the rest of the embryo, and as great
a variety in the transitory shapes of the embryo they produce.
But their end-result is the same : the disposition of the
material for the future internal organs.

These major movements which sweep masses of cells
into the interior are not, however, the sole source of the internal
organs, though they are the sole source of the gut. In most
animals later smaller-scale migrations from the surface contribute a proportion of the tissues which lie between gut and
skin. In particular the nervous system becomes internal by
an independent immigration.

The analysis of how these movements occur is in its
infancy. Orderly changes in cell-shape, probably often due
to cells creeping over each other, through reactions of their
cell-surfaces (which it will be remembered cleavage creates)
seem to be largely responsible. But expansion of the future
outermost layer of the embryo is apparently independent of
the existence of cell-surfaces, since it occurs in the eggs with
suppressed cleavage previously referred to, and it has similarly
been reported in uncleaved frogs' eggs.

What are we to make of the significance of this phase
for the course of development ? It is clear that the disposition
of the material which is going to form particular future
organs is quite different at the end of cleavage from its disposition at the end of gastrulation. The disposition at the
end of cleavage presumably to some extent represents the
arrangement of the developmental mechanisms, which up
to this stage must work in this arrangement to prepare the
future fate of the constituent cells. It is an arrangement
which seems to involve the embryo surface to an important
extent, judging by the characteristic direction of gastrulation
movements from the surface inwards, and it is interesting
that we have experimental evidence that the superficial layer
of the egg is of great importance for the developmental

mechanism, though we have not the faintest idea how it is organized. But the arrangement of the development-determining mechanism of the egg and cleavage stages is the wrong arrangement for the adult; it results, for instance, in the gut material being placed inside out on the surface of the embryo ; and it is completely reorganized by the gastrulation movements. Surprisingly enough, though there is little visible sign of the adult structure, at the end of gastrulation the material which will form the various organs has approximately the same arrangement as it will have in the adult.

Organ-formation

The nearer we get to the adult the more animals of different species differ, and now we are seriously concerned in the particularities which distinguish one kind of animal from another. Generalization becomes increasingly difficult. For safety's sake let us use examples rather than generalities. Consider what happens to a vertebrate, say a frog or a newt (useless animals in many ways, but vital embryological material ; their existence is justified as long as they lay eggs for embryologists).

In such an animal cleavage produces a hollow sphere of cells, the walls of which are several cells thick. At the end of gastrulation we have what is roughly a double-layered sac, the inner layer having formed by the dimpling inwards of the wall of the hollow sphere formed by cleavage. The cavity within the inner wall, which is the future cavity of the gut, is excentrically placed because of a mass of yolk on its ventral side. Quite suddenly, what we have been long awaiting happens. The organs of the animal appear, admittedly only in roughest outline ; seeming to crystallize out in approximately their adult positions. We attain a recognizable if a rather distorted sketch of a real adult animal. The formerly smooth homogeneous sheets of cells, which had streamed into position during gastrulation, break up into separate masses, each a future organ. It is as if the cells of each mass strongly attract each other and clump tightly together, often squeezing each other into new shapes in the process ; while the cells of different but contiguous masses fail to stick, so that gaps appear between them. Such specific attraction and repulsion is indeed probably at the basis of the blocking-out of organs, as Holtfreter's beautiful experiments show. He isolated small groups of cells of different kinds and introduced them to each other in all sorts of combinations, watching their mutual behaviour. Attraction or repulsion depended on who met whom.

Now that the main organization of the future adult is deployed, the rest is a matter of filling in the details. The dramatic period of sweeping change is over, and the task of converting the formless egg into a recognizable animal has been fulfilled. But filling in the details is what really takes the time in development. We have, for instance, so far covered in this account only about the first month of a human being's existence. Nor are the details any less fascinating to the embryologist. The reason I must pay little attention to them is merely that they are so diverse. But it is perhaps worth noting the fundamental cellular processes, various combinations of which are responsible for all the diversity of organ formation.

In the first place, many of the future organs are not yet properly in their definitive place. So migrations and foldings still go on. The turmoil, however, is no longer total as during gastrulation, but sporadic. Then there is still a great deal of cell-division, also sporadic as it was in the phase of gastrulation ; and this cell-division is now accompanied by growth of the daughter-cells so that the products have an increased mass, instead of simply resulting in smaller cells as it does during cleavage and often during gastrulation. The material for this growth comes from digestion of the yolk, or from maternal fluids. Expansion and change of shape are therefore produced by localized mitosis within an organ. Curiously enough there is also quite a lot of cell-destruction, which may remove temporary scaffolding to make way for migration or growth of other tissues, or may separate initially joined structures. Most significant of all, this is the period when the embryonic cells start to lose their hitherto simple form, and begin to assume the shapes and structures which go with their adult functions. For instance, muscle cells elongate and develop contractile fibrils in their cytoplasm; nerve-cells produce long processes and become able to conduct impulses. Furthermore, intercellular substances are secreted in the tissues between gut and epidermis, especially the system of tough collagen fibres of " connective tissue " which is so important in holding the embryo (and the adult) firmly together. The organism begins to assume the functions of an adult.

By a variety of combinations of such processes the organ systems of all animals are blocked out and perfected. Here is the place, however, where we must admit that what all this development has been leading up to is often not an adult but a larva. A larva is a stage of development which for a time leads an active life quite unlike that lived by the adult. It is particularly common in the development of marine and fresh-water animals. It is a break in the steady continuity of development, because before development towards the adult is resumed the larva is partly, often very largely, destroyed by degeneration of its tissues (the process of metamorphosis). The metamorphosis from tadpole to frog, in which for instance tail and gills are lost, is a very mild example. Nevertheless with ups and downs, and at various rates in different organs and in different organisms, the embryonic structure moves ever more slowly to that of the adult. Not that the structure of the adult is really a stable end-point. More slowly still it is continuing the process of change, adjusting to new stresses, ageing away to death.

Evolution has been well-described as a process of adaptation to an ever-increasing severity in the environment, and the more adaptable animals and plants have naturally succeeded best. Now this adaptability depends, as Dr. Abercrombie has said, to a large extent on cross-breeding. This is ensured in most animals by the device of separate sexes. But in plants, with a few exceptions like asparagus and the date-palm, the male and female reproductive organs are either enclosed in the same flower, or are in different flowers on the same plant. Numerous devices have therefore been evolved not only to promote cross-fertilization, but also to prevent self-fertilization either completely or for a time. Every fruit-grower knows that apples are usually self-sterile. The pollen cannot be prevented from falling on the stigma, but the grains from that flower cannot grow right down to the ovary to fertilize the seeds, and unless cross-fertilization has occurred from another variety no fruit will " set ". Such a physiological barrier to fertilization is known as " incompatibility ". *And what of the devices to promote cross-fertilization?* All the beauty of flowers, their coloured petals, their scent, and the nectar they secrete, are just so many lures to attract insect visitors and so ensure that the pollen from one flower will be carried to another. Here is an extract from a famous book published in 1793 and called *The Secret of Nature discovered in the Structure and Fertilization of Flowers*. It was written by a pioneer in the study of the inter-relations of plant and insect life, a gentle Pomeranian priest who was ejected from his living because he " neglected his flock in favour of flowers ".

CHRISTIAN KONRAD SPRENGEL

FLOWERS AND THEIR VISITORS

IN the summer of 1787, as I was attentively considering the flower of the wood cranesbill, I observed that the lower portion of the petals was covered on the inner surface and along both borders with soft, fine hairs. Convinced that the wise Creator of Nature had not caused a single hair to grow without a definite purpose, I reflected long as to what end these hairs might serve. It soon became clear to me that, if it could be assumed that the five drops of nectar secreted by five little glands were designed for the nourishment of some particular insect, it was by no means improbable that some arrangement had been made to prevent the nectar from being spoiled by rain, and that this was the end served by these hairs. . . . Each drop of nectar lies on its gland directly under the hairs on the borders of the two nearest petals. As the flower stands erect and is comparatively large, some raindrops must inevitably fall into it. But none of these can reach the nectar and mingle with it, because they are caught up by the projecting hairs, just as a drop of sweat from the brow of a human being is caught up by the eyebrows or eyelashes, and prevented from flowing into the eye. An insect, however, will be in no wise hindered by these hairs from reaching the nectar.

I afterwards examined other flowers and found that divers of them showed something in their structures which seemed to serve the same purpose. The longer I continued my observations, the more convinced I became that flowers which contain nectar are so disposed that insects can easily reach the nectar, but that rain is intercepted. I concluded from this that the nectar of these flowers is produced chiefly for the sake of the insects, and that it is protected from the rain that they may enjoy it pure and undefiled.

In the following summer I examined the forget-me-not, and found that the flower secretes nectar, and that the nectar is protected from rain. At the same time my attention was attracted by the yellow ring round the opening of the flower-tube, which contrasts so beautifully with the sky-blue petals. Can this too, I asked myself, have some relation to the insects ? Has Nature coloured this ring so brightly that it may show the insect the way to the nectaries ? I examined other flowers in the light of this hypothesis and found that many of them supported it. For I saw that those flowers in which the corona shows differences in colouring have the splashes, figures, lines, or dots at the entrance to the nectaries. Then I reasoned from the part to the whole. If, I reflected, the corona has a particular colouring at one spot for the sake of the insect, then the whole flower is coloured with some reference to it, and if the special colouring of a part serves the insect that has settled on a flower as a guide to the nectar within, then the colour of the whole corona proclaims the flower from afar as a store of nectar to the innumerable insects flying about in search of food.

In the summer of 1789 I examined several species of iris, and found that these flowers could be fertilized in no other way than by insects. Then I searched for other flowers whose structure should show the same thing. My studies convinced me that many, perhaps all of the flowers which secrete nectar are fertilized by insects, and that therefore the nourishing of the insects is an end in relation to them, but in relation to the flower is only a means to an end, which is fertilization, and that the whole structure of such flowers can be interpreted if we bear in mind two points : (1) These flowers must be fertilized by a species, or by several species of insect. (2) That this may be brought about, it is essential that the insects, when they alight on, or creep into the flower, in search of nectar, should rub the pollen off the anthers and deposit it on the stigma, which, to facilitate this, is covered with fine hairs or with a sticky moisture. . . .

He who simply sends out to field or garden to fetch flowers which he examines in his study will never discover Nature's plan in their structure. We must observe them in their natural position, and must note especially whether they are visited by insects, and if so, by what kinds, how these behave, whether, when creeping into the flower in search of nectar, they come in contact with the anthers and stigma, and whether they cause a change of appearance in any part of the plant. In short, we must endeavour to catch Nature in the act.

We must observe the flowers at different hours, and note whether they are day or night flowers, and in different states of weather, as during and after rain, to note how the nectar is protected. But it is especially in the hot noonday hours, when the sun stands high in the cloudless heavens, that we must be diligent in our observations, for it is then that the day flowers appear in all their beauty, and compete with all their charms for the visits of the insects. And it is then that the heat-loving insects are most active on and within the flowers, in pursuit of their own object of revelling in the nectar, while at the same time fulfilling Nature's purpose of fertilizing the flower. In the kingdom of Flora, whose wisdom is not less wonderful than her beauty, there take place, during these noontide hours, miracles of which the mere laboratory botanist does not even dream.

Dr. Huxley's lecture explained briefly how Mendelism fitted in with the idea of evolution through selection, and a further glance at this important matter must now be taken. Its importance will quickly become obvious when we begin, as we very shortly shall do, to consider the implications of what we have learnt for the human race, and notably for the ideals of eugenics.

What is the modern view of the machinery of evolution? Our readers will already have some idea of this, and now Dr. E. B. Ford, Lecturer in Genetics at Oxford, will help them to form a clearer view. We quote an extract from the essay which he contributed to the congratulatory volume presented to Professor Goodrich of Oxford on his seventieth birthday. The latter part of this essay has been omitted as rather too technical, but towards the end of it Dr. Ford quotes Professor Goodrich's opinion on the vexed question of the inheritance of acquired characters : " No single part or character is completely ' acquired ', or due to inheritance alone. Every character is the product both of factors of inheritance and of environment, and can only be reproduced when both are present. Characters are due to responses, and have to be made anew at every generation. Only those characters reappear regularly in successive generations which depend for their development on stimuli always present in the normal environment. Others, depending on a new or occasional stimulus, do not reappear in the next generation unless the stimulus is present." These words should be remembered when any discussion arises about Soviet genetics.

EDMUND BRISCOE FORD

THE GENETIC BASIS OF ADAPTATION

GENETICS, being a subject of recent development, is not only growing rapidly, but changing. It is valuable, therefore, to pause now and then and review the position which some of its concepts have reached. The present occasion provides an opportunity for doing so. It may be used briefly to survey a few aspects of variation, with special reference to selection and, in particular, to the evolution of adaptations. . . .

[Dr. Ford went on to outline certain " fundamental principles " which have already been mentioned in this volume.]

Genetic factors have multiple effects, and they interact with one another to produce the characters for which they are responsible. The chances, therefore, that any random mutation shall upset rather than improve the adjustment of the organism to its environment are very great. However, a gene having beneficial effects will be produced now and then. It will be spread through the population by selection, until its former normal allelomorph has become a rarity : favourable forms will not long remain " varieties ". Consequently, if the mutations which we encounter in genetic work really represent the type of change used in evolutionary progress, we should expect their results almost always to be disadvantageous, as in fact we find they are.

It seems that such " mutant genes " are generally pressed into the recessive state by selection operating against their heterozygous effects. That normal individuals carry them in numbers is now well known : so, too, is the reality of the selective influence brought to bear upon them. . . .

We may reasonably anticipate that genes responsible for minute advantages are constantly spreading in the population, displacing their allelomorphs, and adapting their possessors more perfectly to their environment. The greater variability of abundant as compared with rare species, apprehended by Darwin, provides evidence for this conclusion.

Suppose an advantageous mutation of any kind occurs in one individual in ten million. If this number only breed, the new gene is quite likely to be lost by random extinction, but less likely if a hundred million breed. Furthermore, it can be shown that mutations producing genes of nearly neutral survival value must be excessively rare. Nor can we suppose that they occur with different absolute rates in common and scarce species. Fisher has therefore pointed out that the number of these " neutral genes " which any species can maintain will be nearly proportional to the logarithm of its population. That is to say, when very rare, their existence will favour but little the variance of the commoner forms which, however, can keep more of such genes in reserve. Changes in the environment may from time to time cause the effects of any of them to become slightly advantageous. The gene in question will then spread, giving rise to increased variability, reaching a maximum when it and its allelomorph are present in the population in equal numbers. Consequently, any observable difference in variability, when ascribable to the population level of the species concerned, will be due to the spread of advantageous genes actually engaged in bringing about evolutionary change. Such comparisons are laborious and not easily made. However, two sets of observations have now demonstrated that abundant species are more variable than rare ones. In the first, the colour-variation of a group of night-flying moths was related to the abundance of the species; while in the second, egg-size in birds was selected for similar treatment. The data so obtained demonstrate the existence and spread of genes giving rise to small advantageous effects in neutral conditions.

It has already been pointed out that the influence of a gene may be disadvantageous in one environment and advantageous in another. It is worth while to reflect upon the adjustment of an organism to its surroundings from this point of view. Any form will endeavour to secure optimum conditions, and in these it will remain so far as possible. There is a tendency, therefore, for all species to live in a constant environment, being that which suits them best. This is the one in which they are selected, and in which the effects of the genes are usually judged. In it, too, the gene-complex is balanced so as to enhance favourable qualities and to minimize undesirable ones. The chances are very small that any random change shall prove beneficial in circumstances such as these. As the optimum is left, two components have to be considered. First, the species finds itself in a situation to which it is not properly adapted, and secondly it may not retain its former characters unmodified, for the genes may interact with their fresh surroundings in new ways. The possibility of a random change being useful is a little greater now : still more so when lethal conditions are approached.

It would appear to be an ideal system, therefore, for organisms to vary less in their optimum environment than in any other. There are indications that such an arrangement is sometimes actually realized.

An example of it seems to be provided by certain of the instances in which asexual and sexual methods of reproduction may succeed one another in the Protozoa. For example, it is well known that the asexual type is usually employed by *Paramecium*, the individuals generally dividing by binary fission so long as the culture conditions remain good. In these circumstances, genetic variation is largely restricted to mutation. As the supply of suitable food becomes exhausted and waste products accumulate, conjugation, which normally occurs sporadically, takes place more frequently. It has been shown that this gives rise to a great outburst of genetic variation. Most of the forms produced are less well adapted to the environment than was the original clone. Many, indeed, are visibly deformed, and the majority perish in the intense selection to which they are subjected. On the other hand, this preserves those which can withstand the conditions best, and a temporary improvement in the viability of the stock often takes place. However, a steady deterioration in the culture will of course ultimately lead to an environment which the species is quite unable to tolerate.

In examples such as these we find that variation is reduced when conditions are favourable, for then it will almost certainly lead to nothing but wastage. When they depart from the optimum the organism varies more, by undertaking a greater amount of sexual reproduction. The advantage of doing so at that time is apparent from the improvement which sometimes results.

Now, as any evolving line becomes more highly specialized, the variations which could possibly be of use to it are progressively restricted. Finally, it attains a state of " orthogenesis ", in which the only changes open to the species are those which push it further along the path which it has already pursued. Evolution of this kind is very common, and it must necessarily lead to extinction ; for it is impossible for such forms to adapt themselves in new ways to suit changes in the environment. Their last phase, when almost all variation is bound to be dangerous, should, where practicable, be accompanied by parthenogenesis.

In such circumstances, the process is of rather gloomy significance. When combined with the sexual method, however, it may give rise to very satisfactory results, as it seems to do in many insects. These are akin to the alternation of asexual and sexual reproduction in the Protozoa. As pointed out by Snell, facultative parthenogenesis combined with haploidy is to be regarded as a eugenic measure of a rather special kind. Thus in the honey-bee, all recessive characters are expressed in the males, which arise from haploid eggs : a fact of course uninfluenced by the subsequent chromosome doubling which occurs in the majority of their tissues. Selection will therefore purge the species of genes having disadvantageous effects : while the competitive nuptial flight must make the process an especially rigorous one.

Natural selection means, in a classic phrase, " the survival of the fittest ". So long as nature is left alone this process continues to the benefit of the race. But, so far as mankind is concerned, the one thing which happens in every civilized community is that nature is *not* left alone. In many ways, and from a variety of social and humanitarian considerations, we obstruct the working of natural selection. Are we always wise? *How far does our care for the present generation improve or damage the prospects for future generations?*

In the following lecture Lord Dawson, who was Physician-in-Ordinary to both King George V and King George VI, gives a carefully worded answer to this question.

VISCOUNT DAWSON OF PENN

MEDICAL SCIENCE AND SOCIAL PROGRESS

IN its pursuit of knowledge as such, medicine has no concern for aught else but that pursuit. On the other hand, in the applications of that knowledge it has contacts with an increasing range of sciences and human activities. It is imperfectly realized even now how vast was the change when the medical profession entered the field of preventive medicine. Its responsibilities became not only enlarged but different, and are now concerned in composing the ills not only of individuals but of communities, and not only after but before they have happened.

A vast fabric has grown up, partly medical, partly economic, for protecting and promoting the welfare of the individual from the time of birth to a lengthening old age. The community cares for expectant mothers, for infants, and for school children, who are both fed and educated, and there are services concerned with health, unemployment, widowhood, and old age, to mention but a few examples of this wide embracement of nurture. It is well from time to time to inquire how far this policy bodes well for the future—whether in the desire to help our fellow citizens to-day we are unwittingly damaging the citizens of to-morrow.

In a primitive society the law of natural selection operated, and, speaking generally, the strong survived and the weak were eliminated. As society has developed, the sway of natural selection has lessened. To-day the progress of medical science, the qualities of justice and mercy, and the organized efforts of the community are bringing it to naught. And so the " unfit " are preserved. Nature's method of securing quality of population involved a high death-rate, as well as a high birth-rate. In this way there was a wide though rough selection. If we displace Nature's method of a high death-rate we must secure quality of population in another way.

We hear much of economic " planning " to secure the greater welfare of the people. What will this avail unless we give a like attention to " planning " for a healthy race ? In fact, the fall in the death-rate of infants between 1870 and 1934 from 156 to 60 and of children under 5 from 68 to 18 per 1000 resulted from the quickening of the public conscience and growing medical knowledge. It represented a higher standard of social conduct wholly commendable, and contrasted with the days when deaths of children were regarded as inevitable and even a relief, as when a patient in hospital, on being asked how many children she had had, replied, " Had twelve, buried six ".

Meanwhile the collateral effects of this big fall in the death-rate received but scanty attention. Biology was forgotten. Yet it is now obvious that the former level of the birth-rate could not be maintained with such falling in the death-rate, because families would thus be too large for maintenance. It is one thing to have six children survive out of twelve births and another to have ten survive. This is the economic reason for the coming of birth control.

There are further biological reasons worth considering. During the procreative part of her life, woman has twelve sex cycles a year, and has therefore an oft-recurring proneness to conception. Side by side with this fact place another. From a period antedating history the love of man and woman has involved periodic physical expression. Health and harmony are promoted thereby. Indeed, the organizing of the home and monogamy itself envisage this recurring physical expression, unless the latter is inhibited by the unfitness of either party. In former days this periodic physical expression resulted in high conception-rates (that is, live births plus miscarriages), the results of the number of births being mitigated by high death-rates. With the steady and considerable fall in the death-rates something had to happen : hence contraception. The only alternative was organized abstention, a biological and social revolution as impracticable as it would have been harmful.

What does this falling death-rate portend ? Surely that, however necessary, however right as part of a social policy of betterment, it involves the preservation not only of " good lives ", but also of biologically " bad lives", and the latter, apart from their liability to become an economic burden, yield a high proportion of bad breeders and inefficient parents. Sir George Newman says : " We must face the fact that the enormous reduction in infant mortality has led to increased longevity of a certain number of children who may ultimately prove to be ' unfit ' both physically and mentally ".

Quality of the stock—the promotion of the good and the elimination of the bad—calls for closer attention, and with a low birth-rate is of transcendent importance.

But while there is readiness to discuss and practise " planning " in the sphere of economics, industry, and social service in order to secure the welfare of those now living, there is a shyness—even a shirking—towards the study and determination of the conditions which will produce high quality of children yet unborn. In stressing the importance of environment, the potency of the inborn factor is being forgotten. Yet the strength of hereditary qualities and defects is so well known as to need no enforcement.

This leaving of biological problems in the shade is unfortunate. The facts compel our attention. Regarding the unfit, there come to mind forms of physical defectiveness and disease, mental deficiency, mental disease, and inveterate delinquency. Apart from the hereditary taint which these sometimes carry, such people are undesirable—nay, damaging —as parents for the future generation. Though it is neither desirable nor practicable to interfere with choice as the basis of human mating, we should at least control the propagation of unfit progeny. In this connexion the problem of mental deficiency looms large, because of the numbers of its victims (and they are increasing), the difficulty of its control, and the unfruitful expenditure it involves.

A Board of Education report states : " We spend something like six times more money in the residential upbringing of the mentally defective child than on the upbringing of the normal child in the elementary school ".

Yet the results are indifferent and disappointing. Training cannot raise appreciably and permanently the standard of intelligence in the defective. Nurture needs to have greater regard for biological facts. " For of thorns men do not gather figs."

Civilization having in large measure nullified Nature's method of eliminating the unfit, needs to replace those methods by social determination, by control of environment and inheritance—namely, (1) by methods of nurture to build up fit citizens, *having regard to their biological capacities* ; (2) by means whereby the " unfit " can prevent damage to the community by their production of bad stock and bad homes. . . . Environment should be favourable enough to enable the individual to struggle with reasonable chance of success, but not so favourable as to minimize the need for struggle. The former favours the survival of the fit, the latter of the unfit.

Next, to consider subnormal children, that large number of deficients of varying grades able to live in the community. These should be classified according to their fundamental capacities, having in mind their welfare but also the welfare of the community of to-day and to-morrow. To raise a (mental) deficient by specialized institutional training to a higher efficiency and sensibility, however well intentioned, is of doubtful kindness to him and of disservice to future generations. If retained within the institution, would not a simple kindly shelter more in keeping with his fundamental incapacity bring a greater content without a wasteful expenditure of money on useless forms of education ? If set free in society, with its competitive struggles, aspirations aroused come to naught, performance fails, and the victim gravitates into the " social failure " group, marries within that group, and has children ill-starred from birth onwards. And so the grave problem of the unfit becomes aggravated with the years. Of children born of mentally defective parents nearly half will be subnormal.

Sterilization.—I pass to consider the second heading : means whereby the unfit can prevent damage to the community by their production of bad stock and by bad homes. Sterilization is such an endeavour. Its purpose would be to prevent certain of the unfit from propagating, because either they transmit unfitness to their offspring or they are parents so unworthy as to be unable to give their children a fair chance. It is our aim to prevent disease and disability in our own generation ; why should we not aim to prevent them in the generations to follow ? Sterilization would be a form of preventive treatment in the interests of the race. It is hardly necessary to remind this audience that sterilization imposes no change in sex life. On the other hand, for the suitable cases it brings relief because it guards them from responsibilities for which they are unsuited. It is a branch of therapy comparable to preventive inoculation ; its accomplishment involves in trained hands a simple procedure, and it has no ill

results on health.   Individuals to be considered for this
treatment would be limited to those within the sex-active
period of life, and, as far as mental deficiency is concerned, to
those outside institutional provision.

To form an opinion as to the effectiveness of sterilization
it is necessary briefly to consider the mechanism of inheritance.
The ways and means of inheritance were first unfolded by the
genius of Mendel, but they obtained little notice for many
years until they were rediscovered by De Vries in 1900.
Since then, discovery has been extending, certain conclusions
are assured, and those which are elementary to my theme I
will make bold to bring before you in brief terms.

A new life starts as a single cell within which is a single
nucleus.   This nucleus contains numerous minute particles
which at the pertinent time may be represented diagrammatic-
ally as strings of beads.   The beads are known as genes and
the strings as chromosomes. Each gene is distinctive in
substance and function, and has responsibility for specific
characteristics of body or mind, and further has its allotted
place in the string or chromosome.   If any gene is damaged,
the corresponding characteristic in the owner will be affected.
With fertilization the egg cell contains two strings of genes,
one from each parent, genes of corresponding rôles being side
by side.   Each string would *alone* produce an individual of a
particular pattern.   The pattern of characteristics, bodily and
mental, in fact produced—that is, the new individual—is the
result (the blend) of the influences of the corresponding *pairs*
of genes (paternal and maternal) ; and these genes will be
present in all the cells which result from the division of the
original germ cell.

The genes in the two strings or chromosomes opposite
each other constitute pairs, and each gene of a pair has concern
with the same attribute of body or mind—for example, the
shape of the nose, the colour of the eyes, the quality of intelligence—features of bodily and mental character.   Though
the genes in a pair have the same function, they do not possess

the same influence ; the maternal gene of the pair may make for defective intelligence and paternal for good intelligence. Another gene may represent blue eyes, say of the mother, and the corresponding paternal gene stand for brown eyes. Here is an interesting fact : should one gene of a pair be normal and the other defective, in the child the function is normal— the characteristics follow the superior parent. When the normal gene of the pair dominates the position the resulting characteristic in the child is called " dominant," whereas the evil though hidden influence of the defective gene is called " recessive." With dominant and recessive genes equal in numbers, progeny will show the dominant characters three times as often as the recessive ones. For the child to be defective in some particular of mind or body *both* genes of the corresponding pair must be defective. The malign influence must be double-barrelled. If only one gene is defective and the other gene of the pair normal, the child is normal, though it is in that respect a " carrier " of a defective gene. . . .

This brief exposition of the genes received from parents and passed to progeny, as the machinery of heredity, is, I hope, clear, but it does not pretend to be complete. In the formation of an individual the numerous genes are not water-tight in their functions ; each characteristic feature is the result of the interaction of many genes. The final make-up of the organism is the result of heredity and environment in conjunction, nor can either be excluded from the planning needed to secure a race of high quality. It must not be forgotten that at birth heredity has had nine months' start before environment can have much say. During growth from fertilized egg to new-born babe the characters of this individual are laid down, and the multiple of mere weight is 300 millions.

If one considers hereditary bodily or mental defects—for example, deformed fingers, certain forms of blindness, deafness, chorea, epilepsy, and mental deficiency—in the light of the foregoing brief statement, it becomes apparent (let us exemplify by mental deficiency) that with both genes of the pair damaged the individual is not only personally defective, but will be very prone to transmit mental defect to succeeding generations. With only one gene of the pair damaged the owner will not be personally defective, but will be a carrier of mental defect, which will remain submerged in the strain until there is future matching with a like carrier, when a " full-blown " mentally deficient descendant will appear. There is no means of detecting carriers, who are many times more numerous than actual defectives, and no thinking person supposes that sterilization would touch this part of the problem. It is not, however, an unreasonable hope that some day means for detecting " carriers " may be discovered.

On the other hand, mental defectives are a demonstrable danger, and especially those suffering from milder degrees of deficiency, who are able to live in the community : not only are they unfit to rear children, but if they do perpetuate their strain more mental deficients will be the result. An example will make this clear. A woman's father and mother were within the normal, but they must have been carriers, for the paternal grandfather was feeble-minded, two great-uncles were insane, and a maternal uncle was epileptic. The woman had two illegitimate children, both of whom died of convulsions. She then had fourteen children born in wedlock ; of these, three died, nine were mental defectives, and one son and one daughter appeared normal, but were probably carriers. How long would it be before 1000 defectives resulted from this strain ? This woman was not incompetent to look after her daily needs. If she had been segregated to prevent further procreation, that would have been a deprivation of individual liberty greater than that involved by sterilization and at greater cost to the State.

But this vitiation of the stock is accentuated by the fact that defectives tend to sink to a low economic stratum, where life's varied unfortunates—physical, mental, and moral—gather together ; marriage occurs within the group, and a host of inferior children are born, and become foci of deterioration.

As quality of race is an essential condition of our social progress, even of our survival, we cannot afford to have these vicious strains breeding in our midst, nor must we forget the unkindness, I almost said the cruelty, to the children born. What chance have they to make good ? In days past this racial situation was helped by a high mortality among these inferior children, but this is much less and decreasingly so now.

Sterilization would be advised, as other forms of treatment, with a sense of responsibility and with suasion. The advice would be accepted, not (with possible exceptions) imposed. Voluntary sterilization would be the better way—the advantages of the treatment would spread from the beneficiaries to their like and kind. Experience shows that there are no after-regrets, and that relief is brought to suffering people, who are thereby helped to be better citizens. Sterilization is a restraint on liberty to procreate, but the law restrains liberty in matters of communal health, and when parents are ignorantly obstructive, public medical services can act forcefully in the interests of children. By similar reasoning, why should not future generations be protected ? This leads to the question of co-operation between legal and medical procedures. In recurring delinquency, especially if entangled with mental defectiveness, sterilization would be a protective measure comparable to a sentence of " preventive detention ", but with the advantage that when out of prison propagation by these undesirables would not take place. . . .

This present-day chapter in our evolution—political, economic, and social—is set on a wider stage than its predecessors, because mechanical invention has made the countries of the world one, if not akin. Its outstanding feature is that organized constructive effort seeks more than hitherto to direct man's progress. Such effort can succeed only if it is comprehensive and has regard not only to environment but to the quality of the man himself as determined before as well as after birth.

Medical science learns daily the lesson that for collectivism, however good, it is thus far and no further. With environment sound, it is the individual man, what he inherits, what he becomes, how he overcomes, that has the first and last words. Each man has his own mould in health and illness. Finally, medical science teaches that social progress depends on regard for the narrow way of knowledge, reason, and reasonableness. Please hold that road and to the exclusion of all others, for along it lies the continued greatness of England.

Too much must not be expected from the application of genetical knowledge to human society. It is easy to take too optimistic a view of the possibilities of racial improvement through such measures as sterilization of the unfit, just as it is easy to take too pessimistic a view about the results of man's widespread interference with natural selection. Let us ask one of the greatest experts (Professor Hogben says, " One of the six living people who know most about heredity") to tell us: *How far can we hope to improve the human race by applied genetics?* Gunnar Dahlberg is Director of the State Institute of Human Genetics in the University of Uppsala, and this answer is taken from his book, *Race, Reason, and Rubbish,* in which he made a critical analysis of the Nazi doctrine of the " Herrenvolk ". In it he has something to say about the vexed question of cousin marriages.

GUNNAR DAHLBERG

GENETICS AND EUGENICS

WHEN we have talked about a population with a stable gene equipment maintained without change in successive generations by fixed fertility and mating at random, nothing said so far need imply denial of the possibility of fortuitous changes. The frequency of a gene can increase to some extent in one generation and diminish in another ; but changes of this sort must cancel out in large populations. It is only in small communities that we have to reckon with appreciable deviations in successive generations. Similarly, when we speak of people marrying at random and producing an average number of children, we do not exclude the possibility that there are many children in one family and few or none in another ; but if the average rate of reproduction in a particular group of individuals with a particular equipment of genes is higher or lower than it is in the community as a whole, successive generations will be different. What then concerns us is to get a more precise measure of how quickly such changes take place, *i.e.* to try to estimate how effectively interference with reproduction produces results. If processes of this kind go on in a population, we speak of *selection*.

When people who have a *dominant* character are completely prevented from reproducing their kind, the gene, and therewith the character, is instantaneously eliminated. This is the explanation of why no severe hereditary diseases, contracted early in life, are inherited as dominants. Persons who suffer from such a disease, for instance an extreme type of idiocy, rarely get children, if at all. In a way this means that the disease destroys itself. Even if occasional reproduction does occur, the character will soon disappear. Suppose that it corresponds to one-tenth of the average fertility in the population. Those who have the character in one generation will be one-tenth as numerous as in the previous one. In the next generation there will be one-hundredth, in the next but one one-thousandth, and so on. Even a moderate difference of fertility will quickly produce a depressive effect on the frequency of a dominant character. If, on the other hand, those who have a dominant characteristic have somewhat more than the average number of children, and those with the corresponding recessive one have less or none, the mechanism of change must be different, because the recessive gene will be latent among a section of individuals with the dominant character ; and such individuals do not encounter any impediments to reproduction.

We now know that, if a recessive character is common

persons in which the gene is latent are relatively less common than if it is rare. For this reason it is easy to see that selection acts in a comparatively effective way on a recessive character which is common, and that it is less effective when it is rare. If we make precise calculations on the assumption that all who have the character are prevented from propagation, we can make a graph of the changes. The graph falls steeply at first, and thereafter approximates more and more slowly to the horizontal. For instance, we find that if a recessive character is present in 25 per cent. of the population, completely effective sterilization of all who have it diminishes its occurrence to 11·11 per cent. in one generation. Gradually in this way it reaches a comparatively low percentage while the effect of sterilization becomes weaker all the time. If a recessive character occurs among 0·1 per cent. or ten in ten thousand, we should need to continue with complete sterilization for ten generations, *i.e.* about three hundred years, in order to diminish the frequency to 0·06 per cent. or six in ten thousand. The situation is therefore as follows. Adverse selection of dominant characters is always effective, and we can always rely on getting a quick effect when selection is directed against *common* recessive characters. The same is true, but to a less extent, of characters transmitted in a more complicated way, *i.e.* those which result when several genes of dissimilar pairs get together.

This aspect of the matter is interesting because the differences of fertility which distinguish contemporary social classes and occupational groups would have an effect by no means negligible, if such groups had very different assemblages of genes. One thing much discussed in this connexion is the low fertility of the prosperous classes. Some say that this low birth rate constitutes a danger to society. They assume that the upper classes are more especially the carriers of genes which promote social effectiveness, intelligence, industry, or high character. The same people often complain that the low fertility of well-connected people points to an egotistic or socially indifferent attitude, and that it is therefore essential to encourage a more idealistic outlook in connexion with production of children. It seems rather odd to make out that the upper classes are too selfish, while also urging that such selfish people should be specially encouraged to reproduce.

On the other hand, it may also be said that having few children is not necessarily due to pronounced egotism but to a stronger sense of responsibility for offspring; and that people may not wish to have large families because of the uncertainty of providing for many children so well as they could provide for a few. Even if they are able to help a large family better than other folks, prudent people may feel convinced that they can look after one or two children better than six or seven. On this statement of the case we may perhaps exonerate the prosperous classes from the charge of exaggerated egotism, and what we have then to decide is whether they really constitute a selection of specially intelligent people. Other things being equal, it is quite clear that intelligent people have greater possibilities of getting on in the world than stupid ones have; but many people who belong to the upper classes have not worked their way up to the positions they hold. They inherit them. Although their ancestors may have done the necessary work for rising in the social scale, it is not at all certain that the offspring of such individuals constitute as good a selection as their forefathers.

We may also ask if the process of social selection is itself a good one, i.e. whether people who have worked their way up in the world are really over-blessed with desirable characteristics. As emphasized already, intelligence must be some advantage; but many other traits also play a part in the individual struggle for social position. A certain lack of scrupulousness may help. The very lack of intelligence can lead to equanimity in critical situations and may therefore propel people slowly upwards in some circumstances. For instance, those who lead the crowd and carry people along with them by force of conviction are seldom intelligent.

To make an impression on the masses calls for the assurance that personal opinions are correct. Only those who have this assurance can talk with enough conviction to get results. Only the stupid can arrive at complete certainty. Only those who are not much above the average can express themselves in a way which the masses understand.

Needless to say, the nature of social processes which determine class mobility are too little understood to entitle us to give a decisive verdict. A number of enquiries have been carried out by means of so-called intelligence tests upon children of different social classes. The differences discovered are comparatively insignificant, and only lead to the conclusion that the average divergence between different social classes is not of very decisive dimensions. With regard to inheritance in general it is important to recognize that traits which play a part in social efficiency are very largely influenced by the social *milieu* itself. They depend partly on early training, school education, and so forth, and we may be sure that the hereditary mechanisms also involved are of a very complicated type. Whatever we may mean by the word intelligence, it is certainly not inherited as a simple mendelian character, but depends on a multiplicity of genes. Both these considerations diminish the likelihood that big differences of gene equipment distinguish different social classes. . . .

We can conclude that the majority of annual recruits to the more prosperous classes come from the same social stratum as their parents, and that only a small number of gifted individuals of lower social position have the chance of rising.

Beyond this we can scarcely go at present. We must carry out comprehensive researches before we can venture to give a more definite verdict about this important question. Meanwhile we should remember that differences of fertility such as those we have discussed are of comparatively recent date. In France they have existed for a few generations ; but in most other countries they are scarcely more than a generation old. Besides, we have reason to believe that family limitation began among the well-to-do and spread to the poorer classes. In Stockholm, and in some other parts of the world, people with low incomes have begun to have less children than those with larger ones. So it is likely that such differences between social classes are ephemeral. As the problem of a declining birth-rate grows and assumes importance, it is probable that differences of fertility between people of different income levels, such as those we have discussed, will disappear.

In so far as it affects smaller groups, the problem of selection is comparatively simple. If a rare character associated with low fertility, or one which is inherently associated with obstacles to reproduction, is inherited as a dominant, it will soon be eliminated. Thus the effect we get from sterilization is proportionate to the number sterilized. For example, if we sterilize half of all people who have a particular dominant character, the occurrence of the character is diminished by one half in the next generation. On the other hand, where we have to deal with recessive characters or traits transmitted in a more complicated way, the results of interfering with reproduction are without practical significance when the character itself is a rare one. This assertion contradicts widely accepted opinion. People ' have worked for sterilization measures with the more or less explicit aim of diminishing the number of defectives in a population.

Perhaps it is not convincing enough to carry out merely theoretical calculations about the effectiveness of selection, as in what has gone before. We do not feel convinced by arithmetic if we cannot control it in practice. So it is reasonable that we should want more tangible evidence. In this connexion the author has emphasized that we already have such evidence in the occurrence of juvenile amaurotic idiocy This is a form of idiocy combined with blindness. The unfortunates who inherit it die about the age of puberty and never have children of their own. About four or five such individuals are born annually in Sweden, and there are between seventy and a hundred of them alive at the moment. Though this character is subject to so-called natural selection more exacting and stern than could be brought about by any law of sterilization, and though selection operates wherever the character exists, the disease has not been eliminated. This may seem a little odd. We should expect that prevention of reproduction among such individuals would have some influence. We should certainly anticipate that it would lead to the total disappearance of the character in the course of millennia. In reality this is not so. The effect is so insignificant that it would not be noticeable in the comparatively short interval of time which corresponds to the duration of human life on earth.

Let us look at the matter in another way. The gene for a recessive defect must have turned up at least once through a mutation. Perhaps the first person who has the gene does not have children. So the gene disappears. It may also happen that he has several children of whom some get a single dose of it. This allows the gene to be transmitted to their offspring, and it may gradually spread throughout the population. When the gene has spread to some extent in this way, two persons in whom it is latent may happen to marry one another. Among their children we should expect that one of every four will actually exhibit the recessive character, two others will carry the gene, and a fourth will lack it altogether. Now if we prevent those who have the character from reproducing their kind, we cannot reasonably hinder both of those who carry it, and if all families consisted of four children and two parents, there would be two carriers among the offspring for every two parents also carriers. So by getting rid of those who have the manifest character, we do not necessarily affect the number of those in which the gene is latent. A recessive gene must have a certain spread before characters begin to crop up at isolated points in the population. Even if we capture these points we can never suppress the gene below the level of frequency which corresponds to that at which this isolated

outcropping of the character occurs. This lower limit is what
the author has called the *lowest heterozygote limit*.

In this connexion it is worth rubbing in the fact that
corresponding reasoning applies even to defects with a more
complicated method of transmission, for instance characters
which depend on two dominant genes which must come
together if it is to show up. Even if we adopt measures for
the sterilization of parents and brothers and sisters of every
defective at its birth, we could not change the situation.
Rare genes could not be completely stamped out. They can
only be pressed down to a certain limit of rarity.

Hereditary defects and inherited diseases are comparatively
rare phenomena. Even if we set the limit at its highest, we
need scarcely reckon with more than a few per cent. of really
defective people in any population. An estimate by the author
for the Swedish population gives 4 per cent., even when the
limit is stretched to the uttermost. If we lump together all
defective individuals we are perhaps tempted to believe that
we could produce a decided diminution because the figure,
though low, is not so low as assumed in the reasoning set
forth above. However, we have to remember that the group
is not homogeneous. It is made up of numerous dissimilar
diseases such as hereditary blindness, deafness, congenital
disabilities, epilepsy, idiocy, mental dullness, etc. With each
of these classes in its turn we may have to reckon with several
different types of transmission. Among the blind, for example,
we can distinguish at least ten types, and there are many quite
different sorts of idiocy or imbecility. For every such sub-
group the effect of sterilization is zero, and the sum of a set of
zeros is also zero. In other words, all these defects lie near to
the lowest limit of heterozygote frequency.

If we doubt whether this conclusion is a reasonable one,
we ought to take into account the fact that such defective
individuals have had little opportunity of reproducing during
the last few thousand years. They have lived on the verge of
hunger and death, dependent on charity which scarcely

existed. They could not procure their own means of subsistence ; and we may judge the prospect of life for them from the fact that the custom of exposing the newly born with any obvious defect persisted till a comparatively high level of culture. Though people could not bring themselves to kill helpless children, they let them die by themselves of cold and hunger, and surely did so with no lightness of heart. More human sentiments are of comparatively recent date ; and still later came kindly treatment of the children of social misfits. In such circumstances the reproduction of defective individuals must have been practically non-existent in the past; and it is certainly very small at the present day. Despite stringent selection, defectives have not been stamped out. So far as we can form a rough opinion, for which there are no exact figures, they exist also among peoples whose social attitude towards the care of the helpless is not highly developed. Even in primitive communities blind, deaf, and crippled offspring are born.

Though we cannot expect to produce any socially significant reduction of the frequency of defects by measures for sterilization, there is no reason why we should not take *individual* precautions when well-founded conclusions lead us to believe that we can prevent the birth of an individual whose life will be a burden because of some blemish or the like. In individual cases we can prevent a misfortune, and since defective individuals may be a quite appreciable economic burden to a small community, there is no reason why we should not do what we can do for the individual, even if it is not very significant from the standpoint of the population as a whole.

How much can we diminish the occurrence of a rare character, such as a congenital defect, by forbidding cousin marriages ? If we could put a stop to a large number of cousin marriages, we could get a big result ; but we can hardly imagine any real population in which there are many cousin marriages to prevent and at the same time any rare characteristics. In real life we may have to deal with large population groups, such as a large city, where really rare genes occur, but if so there are relatively few cousin marriages to stop. If we are concerned with a small one, many cousin marriages occur, but in a small community no gene can remain really rare. This way of looking at the problem may seem difficult at first sight, but not so difficult if we bear in mind the fact that when we speak of a *very rare* character, we usually mean a trait such as *albinism*. According to a British estimate, this has a frequency of about one in twenty thousand. For a recessive gene to show up, the population must contain at least two people in which it is latent ; and only when two such people marry and have children can we get individuals who show the recessive character because they have the recessive gene in duplicate. In a small community of fifty to one hundred individuals the existence of only two carriers does not constitute a very rare phenomenon in the sense defined.

Calculations which are relevant to this question are somewhat complicated, but the upshot is that the degree of rarity of the gene on the one hand, and of cousin marriages on the other, balance one another. So at best we can only reduce the incidence of a simple recessive character by about 15 per cent. if we put a stop to all unions between cousins. . . .

At the same time we have to remember this. Though prohibition of cousin marriages leads to such a result in the next generation, it does not produce a further diminution in succeeding ones. The result is instantaneous ; but there is no continuous gain from one generation to another. Where we have to deal with characters which are not so rare, and are not therefore near the *lowest limit of heterozygote frequency*, the effect obtained is less considerable. The effect is also less considerable when a character is transmitted in a more complicated way. . . .

With reference to inbreeding in human populations the outcome of our enquiry is therefore this. Neither the grade of consanguineous marriages nor the percentage of those which actually occur can be said to have any noteworthy significance.

Only when very rare recessive characters are concerned, or when the method of inheritance is complicated, can the exclusion of cousin marriages have an appreciable effect. Perhaps the reader will ask at this point if it is so great that we really ought to take action by preventing cousin marriage. Before trying to answer this we should remember that we cannot rest content with forbidding some cousins to marry. We have to stop all of them if we hope to get measurable consequences, and we do not in fact know whether a gene is latent in any particular pair of cousins. It is therefore a matter of settling whether healthy people should be compelled to make sacrifices in order to get a diminished frequency of defectives in a later generation. Obviously it is not the province of science to decide how many of us should immolate ourselves on the altar of coming generations.

The conclusion to which we have come has also a brighter side. Since we cannot stamp out rare defects, we cannot stamp out advantageous characters which are rare, especially when inherited in a complicated way. For instance, we cannot get rid of intelligence. People have actually tried it on. Celibacy in the Roman Catholic Church is a trick of this kind. At one time, more gifted individuals chose to take vows, and for many gifted individuals of the poorer classes such a decision was the only opportunity of drinking from the springs of culture. Though celibates were prevented from having children, Roman Catholic countries have not any outstanding lack of gifted individuals. Even if some countries adopt measures specially directed against persons who have sufficient intelligence to form independent opinions in opposition to prevailing superstition, we need not expect that they will be able to prevent the birth of single individuals with more than normal intelligence. Stupid people cannot stamp out genius.

Let us now return to the evolution story and look at some of the
wonderful steps by which progress was achieved. Many of these steps
were concerned with improved methods of internal communications—
the flow of sap in plants, and the blood-stream of animals which carries
food and oxygen to every cell in the body. Systems of internal controls
were also developed, and two of these deserve closer attention. The
earliest, shared by both plants and animals, is the endocrine, that is the
production by certain cells of hormones which stimulate other cells,
to which they are carried by the body fluids. The second, developed
only by animals as a product of their need for immediate perception and
quick movement, is a nerve system. *What is the evolutionary importance
of the nervous system?* An answer to this question is provided in the
first part of a lecture by Sir Edward Sharpey-Schafer, who was Professor of Physiology at Edinburgh for many years, and who played a
leading part in the discovery of the functions of the endocrine glands.

SIR EDWARD ALBERT SHARPEY-SCHAFER

THE HUMAN POST OFFICE

IT is the development of the nervous system, although not
proceeding in all classes along exactly the same lines,
which is the most prominent feature of evolution. By and
through it all impressions reaching the organism from the
outside are translated into contraction or some other form of
cell activity. Its formation has been the means of causing
the complete divergence of the world of animals from the
world of plants, none of which possess any trace of a nervous
system. Plants react, it is true, to external impressions, and
these impressions produce profound changes and even comparatively rapid and energetic movements in parts distant
from the point of application of the stimulus—as in the well-
known instance of the sensitive plant. But the impressions
are in all cases propagated directly from cell to cell—not
through the agency of nerve-fibres ; and in the absence of
anything corresponding to a nervous system it is not possible
to suppose that any plant can ever acquire the least glimmer of

intelligence. In animals, on the other hand, from a slight original modification of certain cells has directly proceeded in the course of evolution the elaborate structure of the nervous system with all its varied and complex functions, which reach their culmination in the workings of the human intellect. " What a piece of work is a man ! How noble in reason ! How infinite in faculty ! In form and moving how express and admirable ! In action how like an angel ! In apprehension how like a god ! "

But lest he be elated with his psychical achievements, let him remember that they are but the result of the acquisition by a few cells in a remote ancestor of a slightly greater tendency to react to an external stimulus, so that these cells were brought into closer touch with the outer world ; while on the other hand, by extending beyond the circumscribed area to which their neighbours remained restricted, they gradually acquired a dominating influence over the rest. These dominating cells become nerve-cells ; and now not only furnish the means for transmission of impressions from one part of the organism to another, but in the progress of time have become the seat of perception and conscious sensation, of the formation and association of ideas, of memory, volition, and all the manifestations of the mind !

The most conspicuous part played by the nervous system in the phenomena of life is that which produces and regulates the general movements of the body—movements brought about by the so-called voluntary muscles. The movements are actually the result of impressions imparted to sensory or afferent nerves at the periphery—e.g. in the skin or in the several organs of special sense ; the effect of these impressions may not be immediate, but can be stored for an indefinite time in certain cells of the nervous system. The regulation of movements—whether they occur instantly after reception of the peripheral impression or result after a certain lapse of time ; whether they are accompanied by conscious sensation or are of a purely reflex and unconscious character—is an intricate process, and the conditions of their co-ordination are of a complex nature involving not merely the causation of contraction of certain muscles, but also the prevention of contraction of others. For our present knowledge of these conditions we are largely indebted to the researches of Professor Sherrington.

A less conspicuous but no less important part played by the nervous system is that by which the contractions of involuntary muscles are regulated. Under normal circumstances these are always independent of consciousness, but their regulation is brought about in much the same way as is that of contractions of voluntary muscles—viz. as the result of impressions received at the periphery. These are transmitted by afferent fibres to the central nervous system, and from the latter other impulses are sent down, mostly along the nerves of the sympathetic or autonomic system of nerves which either stimulate or prevent contraction of the involuntary muscles. Many involuntary muscles have a natural tendency to continuous or rhythmic contraction which is quite independent of the central nervous system ; in this case the effect of impulses received from the latter is merely to increase or diminish the amount of such contraction. An example of this double effect is observed in connexion with the heart, which—although it can contract regularly and rhythmically when cut off from the nervous system and even if removed from the body—is normally stimulated to increased activity by impulses coming from the central nervous system through the sympathetic, or to diminished activity by others coming through the vagus. It is due to the readiness by which the action of the heart is influenced in these opposite ways by the spread of impulses generated during the nerve-storms which we term " emotions " that in the language of poetry, and even of every day, the word " heart " has become synonymous with the emotions themselves.

The involuntary muscle of the arteries has its action

similarly balanced. When its contraction is increased, the size of the vessels is lessened and they deliver less blood ; the parts they supply accordingly become pale in colour. On the other hand, when the contraction is diminished the vessels enlarge and deliver more blood ; the parts which they supply become correspondingly ruddy. These changes in the arteries, like the effects upon the heart, may also be produced under the influence of emotions. Thus " blushing " is a purely physiological phenomenon due to diminished action of the muscular tissue of the arteries, whilst the pallor produced by fright is caused by an increased contraction of that tissue. Apart, however, from these conspicuous effects, there is constantly proceeding a less apparent but not less important balancing action between the two sets of nerve-fibres distributed to heart and blood-vessels; which are influenced in one direction or another by every sensation which we experience and even by impressions of which we may be wholly unconscious, such as those which occur during sleep or anæsthesia, or which affect our otherwise insensitive internal organs.

A further instance of nerve-regulation is seen in secreting glands. Not all glands are thus regulated, at least not directly ; but in those which are, the effects are striking. Their regulation is of the same general nature as that exercised upon involuntary muscle, but it influences the chemical activities of the gland-cells and the outpouring of secretion from them. By means of this regulation a secretion can be produced or arrested, increased or diminished. As with muscle, a suitable balance is in this way maintained, and the activity of the glands is adapted to the requirements of the organism. Most of the digestive glands are thus influenced, as are the skin-glands which secrete sweat. And by the action of the nervous system upon the skin-glands, together with its effect in increasing or diminishing the blood supply to the cutaneous blood-vessels, the temperature of our blood is regulated and is kept at the point best suited for maintenance of the life and activity of the tissues.

The action of the nervous system upon the secretion of glands is strikingly exemplified, as in the case of its action upon the heart and blood-vessels, by the effects of the emotions. Thus an emotion of one kind—such as the anticipation of food—will cause saliva to flow—" the mouth to water "; whereas an emotion of another kind—such as fear or anxiety—will stop the secretion, causing the " tongue to cleave unto the roof of the mouth ", and rendering speech difficult or impossible. Such arrest of the salivary secretion also makes the swallowing of dry food difficult : advantage of this fact is taken in the " ordeal by rice " which used to be employed in the East for the detection of criminals.

The activities of the cells constituting our bodies are controlled, as already mentioned, in another way than through the nervous system, viz. by chemical agents (hormones) circulating in the blood. Many of these are produced by special glandular organs known as internally secreting glands. The ordinary secreting glands pour their secretions on the exterior of the body or on a surface communicating with the exterior ; the internally secreting glands pass the materials which they produce directly into the blood. In this fluid the hormones are carried to distant organs. Their influence upon an organ may be essential to the proper performance of its functions or may be merely ancillary to it. In the former case removal of the internally secreting gland which produces the hormone, or its destruction by disease, may prove fatal to the organism. This is the case with the suprarenal [adrenal] capsules : small glands which are adjacent to the kidneys, although having no physiological connexion with these organs. A Guy's physician, Dr. Addison, in the middle of the last century showed that a certain affection, almost always fatal, since known by his name, is associated with disease of the suprarenal capsules. A short time after this observation a French physiologist, Brown-Séquard, found that animals from which the suprarenal capsules are removed rarely survive the operation for more than a few days. In the concluding decade of the last century interest in these bodies was revived by the discovery that they are constantly yielding to the blood a chemical agent (or hormone) which stimulates the contractions of the heart and arteries and assists in the promotion of every action which is brought about through the sympathetic nervous system (Langley).   In this manner the importance of their integrity has been explained, although we have still much to learn regarding their functions.

Another instance of an internally secreting gland which is essential to life, at least to its maintenance in a normal condition, is the thyroid.   The association of imperfect development or disease of the thyroid with disorders of nutrition and inactivity of the nervous system is well ascertained.   The form of idiocy known as cretinism and the affection termed myxœdema are both associated with deficiency of its secretion : somewhat similar conditions to these are produced by the surgical removal of the gland.   The symptoms are alleviated or cured by the administration of its juice.   On the other hand, enlargement of the thyroid, accompanied by increase of its secretion, produces symptoms of nervous excitation, and similar symptoms are caused by excessive administration of the glandular substance by the mouth.   From these observations it is inferred that the juice contains hormones which help to regulate the nutrition of the body and serve to stimulate the nervous system, for the higher functions of which they appear to be essential.   To quote M. Gley, to whose researches we owe much of our knowledge regarding the functions of this organ : " The origin and use of the highest faculties in man are conditional on the purely chemical action of a product of internal secretion.   Psychologists, ponder on these facts ! "

Sir Edward's lecture has introduced two of the endocrine glands, the thyroid and the adrenals. The thyroid seems to control the steady output of energy for normal, everyday conditions, while the adrenals are the " booster " glands, setting free a sudden extra supply of energy for emergencies. The relative sizes of these glands reflect the habit of life of different species. The cat tribe, from the lion downwards, which seize their prey by a sudden leap or rush, have adrenals much heavier than their thyroid. So have the animals on which they prey. In the elephant they are about the same weight, while in man the thyroid is two and a half times as heavy as the adrenals.

Here is a fuller account of these and other endocrine organs by Professor Zuckerman. *What is a hormone?* Many people have made their first acquaintance with this word " hormone " through using the new selective weed-killers, which so stimulate the growth of certain parts of the weeds that they literally " burst themselves ". Gardeners without any weeds in their lawns may have used a different hormone preparation to promote the setting of fruit, or to stimulate the formation of roots on cuttings. Plant hormones (or auxins as they are sometimes called) are different from animal hormones, but, as Professor Zuckerman points out, all mammals seem to possess the same set of endocrine glands, and to be affected in the same way by the same hormones whether these are self-produced or taken from another animal. Here is another remarkable instance of the unity of life in a large class of living things. No doubt the interplay of the glands has been an important factor in the evolution of mankind, just as it now influences the formation of the individual personality.

SIR SOLLY ZUCKERMAN

WHAT IS A HORMONE?

IT is easier to start by giving you the names of some hormones than to tell you straight away what a hormone is.

The substance called insulin is a hormone. Insulin is used in the treatment of sugar diabetes, which is a disorder that results from the inefficient activity of clumps of certain specialized cells in the pancreas, an organ more commonly known as the sweetbread. Before the discovery of insulin, this disease was, for all practical purposes, incurable. Insulin has transformed it into a manageable disorder.

Thyroid extract is another hormone. It is used to treat

disorders that result from under-functioning of the thyroid gland, which is a small organ that lies in front of the windpipe at the base of the neck. The most extreme of these thyroid disorders are cretinism and myxœdema. Cretins are individuals whose thyroid glands have functioned at too low a level right from early infancy, and whose physical and mental development is as a result considerably retarded. Myxœdema is due to deficient functioning of the thyroid, beginning, as a rule, in middle life. A person suffering from this illness develops a puffy face and dry skin, and rapidly loses hair— and what is perhaps more important, physical and mental energy.

Insulin and thyroid hormone are extracts which chemists manufacture from the sweetbreads and thyroid glands of cattle and sheep. What the doctor does when he treats a patient with these substances is to give back to the body something which, for some reason or another, it is producing too little of, or which it has stopped producing at all. We can start off, then, by saying that hormones are substances produced by special glands, and that the conditions in whose treatment they are mostly used result from disordered activity of these glands.

The known hormone-producing glands are few in number, and very different from the multitude of other glands found in the body. The most obvious point of difference is that ordinary glands discharge their secretions through ducts, and that hormone-producing glands pour their secretions directly into the blood-stream. Examples of ordinary glands are the sweat glands and the salivary glands. The skin contains thousands of clumps of sweat-producing cells, each clump arranged in the form of a tightly coiled tube whose secretion passes through a straight duct to open by a minute pore on the surface of the body. The salivary glands are more complex in character. There are three pairs of these glands. They lie in the tissues which make up the wall of the mouth, and each consists of a mass of lobules of secretory cells so arranged that the secretions of each lobule pass along a minute duct, which joins a neighbouring duct and so on, until, in the case of two of the three pairs of glands, the salivary juice of the entire organ enters the mouth through one main duct. With the help of a mirror one can easily see the openings of two such ducts at the base of the central fold that one finds under one's tongue.

As examples of hormone-producing glands let us consider again the pancreas and thyroid.

The pancreas produces both digestive ferments and the hormone insulin. The bulk of the pancreatic tissue is concerned in the manufacture of digestive ferments, without which we could not break down the food we eat into simpler materials that can be absorbed into the blood-stream. These ferments pass through the usual kind of duct system into the intestine. The cells that produce insulin are scattered in small clumps throughout the pancreas. They have no connexion with the system of ducts through which the digestive ferments pass, and their secretion, insulin, is immediately absorbed into the minute blood vessels by which they are enmeshed.

This dual function of the pancreas is a more complicated arrangement than one finds in the thyroid, which secretes only one product, thyroid hormone. The thyroid has no ducts, and its secretory cells are arranged in the form of a honeycomb. Minute thin-walled blood-vessels pass between the cellular walls of neighbouring compartments. The hormonal secretion of the cells is absorbed into the blood and is also stored in the compartments themselves.

These differences between ordinary and hormonal glands are sometimes characterized by the terms " externally secreting glands " and " internally secreting glands ". The internal secreting or hormone-producing glands are also known as endocrine glands, and the branch of science which is concerned with their special study is called Endocrinology.

Let us imagine that we could remove from a person all the

insulin-producing cells in the pancreas. The result would be that he would lose the power to carry out various steps in the utilization of the food he eats, in particular carbohydrates. In this way we should reproduce the clinical condition of sugar diabetes. If to such an experimentally transformed individual you then gave insulin, he would recover his capacity to handle his food, and would become well again. These experimental procedures would show insulin to be some principle essential for the maintenance of health—and even life—because of the control it exercises over the chemical conversions undergone by the carbohydrate food we eat in its transformation into fuel for body energy.

Supposing we removed the thyroid from an animal. Symptoms corresponding to myxœdema or, if we carried out the experiments on very young animals, those of cretinism, would supervene. If we now gave the experimental animal thyroid extracts, its symptoms of deficiency would disappear. These experiments would show that the thyroid is in some way concerned in the control of those processes which provide the body with energy, and with a number of other functions that we need not refer to here. Further experiment would show that its chief activity is that of regulating the rate at which the foodstuffs we eat are transformed chemically in the body, and then used as providers of energy. Both insulin and thyroid hormone thus appear to be concerned in the control of what are generally called metabolic processes, which are the processes of chemical transformation undergone by our food and by the constituents of the body tissues. A deficiency of either hormone leads to grave defects, but both play very different rôles in the complicated process of food-utilization.

Let us look at an example of a different kind of hormonal action. The sex glands, in addition to producing sperms in the male and egg cells or ova in the female, are also hormone-producing bodies. If one castrates a cockerel, it will not develop as a rooster, but becomes what is called a capon. The colouring of its plumage remains more or less normal, but it does not develop a comb, wattles, or spurs, and it does not crow. If we inject into a capon an extract of testis, the bird rapidly takes on all the external characteristics of a rooster—although, because it cannot produce sperm, it remains infertile. On the other hand, inject into it an extract of ovarian tissue, and the bird's plumage loses the bright colours of the male, and assumes those of the hen.

The active principles in these sex-gland extracts, like that in the thyroid gland and in the pancreas, are specific chemical substances that can often be prepared in the form of pure crystals. Hormones we can therefore define as the chemical products of specialized glandular tissues that are taken up in the blood-stream, and which have far-reaching effects on the processes of the body. Before we consider these effects further, let us enumerate the glands that are concerned.

We have already mentioned the thyroid and pancreas, and also the sex glands. The other known hormone-producing organs are the adrenals, two small structures which rest on the top of the kidneys ; the parathyroids, four minute beads of tissue which lie behind the thyroid ; and the pituitary, which is found within the skull below the base of the brain, and which is little bigger than a sixpenny bit. Biological experiments and chemical research have shown that all these tissues produce specific hormones—some more than one—and that they are all concerned in the regulation of vital functions of the body.

The central part of the adrenal glands produces a substance one of whose effects is to cause the smaller arteries to constrict, with a consequent raising of the blood-pressure of the body. Another of its actions is to cause a rapid shift of sugar from the liver, where it is stored, to the blood-stream, by which it is carried to those tissues—for example, the muscles—which need to burn it in order to obtain the energy to act.

The outer part of the adrenal glands differs in microscopic

structure from the central tissue, and has an entirely different function. It produces chemical substances that are absolutely essential to the well-being of the body. One of their functions is to take part in the control of food metabolism. Another is the control of the water exchange of the body. By weight, about 80 per cent. of our tissues—muscle, skin, brain, or what you will—consists of water. This proportion varies but has to be maintained within narrow limits. It is dependent on the balance between the amount of water we drink, or take in as part of our solid food, and that which we excrete. This balance is largely controlled by the secretions of the outer part of the adrenals.

The parathyroid glands produce a secretion which plays an essential part in the control of the calcium metabolism of the body. Our bones consist very largely of calcium. Under normal conditions calcium is always being deposited in, and removed from, the bones through the intermediary of the blood-stream. The balance between deposition and absorption determines the physical state of our skeletal structure, and this balance is under the control of the secretion of the parathyroids. In parathyroid deficiency, the bones become more dense, and sometimes concretions of calcium are actually deposited in tissues—for example, the brain—in which they are never found normally. When the parathyroids produce too much of their hormone, calcium is removed so fast from the bones that they become porous and frail. In this clinical condition fractures are very common.

The pituitary gland consists of two distinct and functionally different parts. The hinder and smaller part, like the outer zone of the adrenals, is concerned in the control of the water balance of the body. The front and larger part is a very important hormone-producing structure, for upon its proper functioning depends not only the control of a number of general bodily functions, but also the normal behaviour of the other hormone-producing glands of the body. Normal growth depends upon a normal rate of production of one of its

secretions which is usually called " growth hormone ". If
the pituitary produces too much of this principle, and if it
starts doing so in childhood, the body will continue growing
out of all normal proportions. It is a safe bet that most
people taller than, say, 6 feet 6 inches, have suffered in their
infancy from over-activity of the pituitary. All the giants
exhibited in fairs and circuses owe their height to this cause.
If the pituitary begins to produce too much hormone once
adult life has been attained, it does not have this effect, for
it is impossible for the hormone to stimulate growth in the
long bones of the body, such as the thigh and shin-bones,
after they have passed the age at which they normally stop
growing. On the other hand, the excess hormone will act
upon certain other bones and tissues of the mature body,
and they then start showing disproportionate growth. The
parts most affected are the hands and fingers, and the lower
jaw.

   Another special chemical product of the front part of the
pituitary controls the development of the reproductive organs.
Thus, if one experimentally removes the pituitary gland from
an immature animal, not only will its body fail to reach adult
proportions, but its reproductive organs will never mature and
function. Experiment has also shown that the reproductive
organs of immature animals can be matured and made functional as a result of injection of that hormone of the pituitary
which is responsible for the control of the reproductive tract.
Separating the different principles of the pituitary is a very
difficult chemical operation.

   In addition to these two functions, it should be noted, as
I have already indicated, that when the pituitary is removed
from an animal, all its other endocrine organs—for example,
the adrenal glands and the thyroid—diminish in size, and
produce less of their chemical secretions. The pituitary is in
fact the master-gland of the endocrine system. On the other
hand, it is not as vital for the continuation of life as such other
glands as the adrenals or the parathyroids. Thus animals

whose pituitaries have been removed survive very much longer than those which have lost their adrenals or para-thyroids. But whichever of these glands is removed, or spontaneously becomes deficient in its activity, the organism almost immediately becomes very abnormal.

I have done no more than sketch in bold outline some of the more important functions of the hormone-producing glands. What I have told you is sufficient to indicate that hormones are chemical regulators of body functions. The activities of the body are also regulated by the brain and the other parts of the nervous system. Probably all bodily processes have a dual nervous and chemical control. This applies not only to the physical but also to the psychological functions of the body. Our personalities, for example, are very sensitive to changes in the hormone balance of the body.

I have repeatedly implied that hormones are specific chemical substances. We know the exact formula of certain hormones—for example, those produced by the sex glands and by the outer cells of the adrenals. On the other hand, the chemical constitution of the hormones of the pituitary gland has not, as yet, been determined. We know enough, however, to say that the hormones of the pituitary are also specific chemical substances. The problem of their analysis is a very complicated one, for it is no less than that of determining how the atoms are arranged to make a molecule of protein. By " molecule " one means the smallest unit of any given sub-stance that can be defined. By " protein " we mean a class of molecules of which the element nitrogen is a constituent; which make up a large part of the living tissue; and which, in fact, impart to this tissue the capacity of " cellular re-production ". The chemical nature of proteins is one of the central problems of biological chemistry, and we are still far from its solution.

It is of interest that the properties of some of the natural hormones—for example, those of the sex glands—can be duplicated by certain synthetic chemical substances whose molecular structures are very different from those of the natural hormones. It is clear from this fact that specific hormonal action depends upon some characteristic configuration of atoms in a molecule, and not upon the total character of the molecule. We do not know, however, the configuration essential for any single kind of hormone action—any more than we know whether or not we have already discovered all the hormones that operate in the body.

Lastly, there is one general aspect of the action of hormones which I should like to mention. So far as is known, there is no essential difference between the chemical character of a hormone produced in one animal species and that produced by the corresponding tissue in another species. For example, the secretions of a sheep's pituitary gland are just as effective when injected into mice as they are in sheep. As a corollary, it follows that the endocrine glands, which are the same in number and microscopic character throughout all mammalian species, control corresponding functions in all. Furthermore, since the hormones are concerned in the regulation of bodily processes, and in so far as some bodily processes are affected by the action of several hormones, it follows that the activity of all hormones fits into a single integrated pattern. One may suppose that this pattern is self-regulating, and that it is maintained by mutual interaction between the different hormone-producing glands.

We have seen that all mammals produce the same hormones, and all are affected by them in the same way. No doubt the same principle applies to birds, and here we have the key to that phenomenon which has puzzled men ever since those early days when Greek poets and Hebrew prophets pointed out that " the stork in the heaven knoweth her appointed times ; and the turtle and the crane and the swallow observe the time of their coming ". *What makes birds migrate ?*

Sir Landsborough Thomson, a past president of the British Ornithologists' Union, gives a modern answer to this question in the following broadcast talk. It would have been gratifying if we could have followed this up with another lecture on how birds navigate when they perform their long journeys, but this problem is still a matter for heated argument among naturalists. It seems most probable that birds find their way as the early sailors did when they passed out of sight of land—using a sense of time allied with observations of the sun, and remembering the distance travelled, or their " dead reckoning ".

SIR LANDSBOROUGH THOMSON

WHY DO BIRDS MIGRATE?

WE are all familiar with the fact that birds, like other animals, are adapted to their environment. It is characteristic of many kinds of birds, however, that they have two environments, which they inhabit at different seasons, and for different purposes. One of them is the breeding area, and the other may be loosely described as winter quarters.

The migration of birds is a tremendous phenomenon. It is also an outstanding example of animal behaviour, which presents a number of problems of great interest to the biologist. We will consider only one of these at the moment, namely the nature of the stimulus that brings this behaviour into action at the appropriate times—twice a year in the life of every migratory bird. To see this question clearly, we must first get rid of a few possible misconceptions. In the first place, it may be obvious that migration serves a useful purpose in the life of birds, permitting them to exploit the opportunities of two different regions at the seasons most favourable in each. But an event does not take place just because it is advantageous : there must be causative factors which make it happen, and so allow the advantage to be secured.

Again, we cannot suppose that birds migrate, as some human beings do, because of an intelligent awareness of the advantages to be obtained. Even if we could attribute such reasoning powers to birds, we should have to remember that

the winter which they avoid is something unknown to them—
something outside the experience not only of the individual
but of the race for countless generations. (We are talking
here of migration in its most highly developed form.) Further,
the most notable migrants cannot be thought of as compelled
to migrate by the stress of external forces. They are not
driven away by approaching winter. Some birds migrate well
before the summer is over; and some travel much farther
than seems necessary—finding not merely a milder winter
but the winterless climate of the tropics, or even a second
summer (in which, by the way, they do not breed) far in the
southern hemisphere. And we must not forget the return
migration in spring as well as autumn departure.

Where, then, do these negative statements take us?
Clearly to the conclusion that migration is the expression of
an inborn instinct. Let us note at once, however, that this
word "instinct" is itself merely a name for something that
we do not fully understand, but which we know to be widely
present among animals. Leaving aside the more general
question of the nature of instinct, we are in this case faced
with two particular problems. What was it which long ago
implanted the instinct in migratory species of birds and induced
its development during their evolution? And what immediate
stimulus awakens the instinct to active expression on each
occasion? In framing the two questions, we distinguish, so
to speak, between the hand that originally packed the explosive
charge in the cartridge and the hand that pulls the trigger and
so fires the shot. At the moment we must confine our discussion to the second problem—the trigger or stimulus
problems. It is one on which a good deal of new light has
been shed in comparatively recent years, and it may therefore
be fittingly considered under the heading of science progress.

Before the modern evidence was available, the tendency
was to attribute the stimulus to external factors found in the
environment. It was obvious that migration was a seasonal
phenomenon. Moreover, changes much too slight to act

as compelling forces might suffice to pull a delicately adjusted trigger. The possibilities were not limited to climatic changes, but extended to the shortening (or lengthening) period of daylight and to secondary influences exerted through food supply.

We do know that meteorological factors play a part, but they seem to do so mainly in determining the exact day of departure within certain limits. Migration tends to occur when the weather conditions at the starting-point are favourable for flight. So a high barometric pressure is effective at either season of migration, and particularly so if it follows a period of low pressure during which movement has been held up. A falling temperature seems to have some influence in autumn, and a rising temperature in spring. But the important thing to notice is that such weather conditions may occur at any time of year but they have this effect only when they do so near the appropriate dates.

This brings us to the conception of a state of readiness in the bird itself—a physiological and psychological condition in which the trigger is at full cock. Only a slight external stimulus is then required to release the pent-up force : possibly, in some circumstances, a mere increase in the internal tension would suffice by itself. There is observational evidence in favour of this view in the well-known restlessness of birds just before migration, as at the time of the autumn " flocking " of the swallows. This unrest has also been noted in captive birds restrained from migrating. A more tangible sign is a heavy deposition of fat, which does not occur simultaneously in closely related non-migratory species.

The modern contribution is in the form of experimental evidence, the tendency of which is to stress the significance of an internal stimulus with a physiological basis. Definite shape was first given to an explanation on these lines by experiments made by Professor Rowan in Alberta just over twenty years ago. He used juncos, finch-like birds, normally only summer visitors to that part of the world. He showed that these birds could survive in outdoor aviaries during the severe winter if they were well provided with food and shielded from direct blizzard. He subjected some of them in the late autumn to periods of artificial lighting which had the effect of gradually lengthening their days at a time when the natural days were shortening. The birds so treated, unlike the controls, were brought back into brooding condition, which they would not naturally have attained till the following spring.

This observation opened up a new line of research in reproductive physiology, but the significant point for us just now is the effect of the experimental treatment on migratory behaviour. Some of the control birds—those whose day had not been artificially lengthened—were liberated in winter but they remained near the aviaries, their urge to migrate having apparently passed. But when birds in process of being brought back to breeding condition were liberated, they disappeared, and were presumed to have at least attempted migration. It is worth noting in passing that this effect was not produced by ultra-violet light, the extra daily ration being provided by ordinary electric bulbs. And in a subsequent experiment the same effect was produced when the artificial lengthening of the days took the form of periods of enforced exercise in the dark, exercise produced by a moving bar in the aviary. It was therefore not the length of daylight as such that counted, but the duration of bodily activity. This work of Rowan has been followed up by other experiments. These have, in the main, confirmed his results and have also extended them in various ways. Wolfson in California, for instance, has worked with a related species of junco, which winters there, and has shown that migration in spring can be induced by similar treatment two months before the normal time.

On the other hand, as so often happens, further investigation has revealed new complexities. The interpretation is not so simple as at first appeared. According to Rowan's original view, readiness to migrate occurred when the reproductive organs were either increasing or decreasing in activity, and was in abeyance when these were at either their maximum or their minimum. There are now some results at variance with this, and in any event it fails to explain migration by young birds in their first autumn. The more probable view is that both reproductive state and migratory readiness are controlled by the activity of the pituitary gland, rather than that the one depends directly on the other. The pituitary in its turn is doubtless influenced by the brain, reacting to external factors perceived through the senses. But whatever the precise mechanism may be, it still seems certain that readiness to migrate is a definite physiological state which reflects certain environmental influences of a seasonal character.

Now let us return for a moment from the laboratory to the field, and consider these external influences a little further. Shortening days in autumn and lengthening days in spring seem to be the most probable form of the primary stimulus. Certainly no other seasonal factor would accord so well with the proverbial punctuality of migration. Within narrow limits, after the primary stimulus has induced the state of readiness, a secondary stimulus in the form of meteorological conditions may determine the exact date of departure. But once again this explanation is probably too simple—or at least it will not fit every case without modification. As framed, it applies to birds which breed, say, in northern Europe, and winter in the Mediterranean area, living wholly within arctic and north temperate latitudes. But what of the bird which winters—that is to say, passes its off season—on the equator, where there is no change in the length of day? Or of the bird which crosses into the southern hemisphere and starts what is really its spring migration when the days are shortening there and all the local conditions are those of autumn? There are even migrants which live and travel entirely within the tropics.

The possible explanation is that the annual cycle of physiological change in the life of the bird, both reproductive and migrational, is in some circumstances attuned to seasonal influences in the environment other than changing length of day. It is conceivable, also, that there is an inherent rhythm in this cycle, and that in a relatively stable environment the periodicity may reveal itself with the mere passage of time. That, then, is where this problem stands at the present moment. We can at least claim that, largely as the result of experiment, considerable progress has been made towards a better understanding of what is involved.

In conclusion, we should remember that this is merely one of the scientific problems presented by bird migration. However near we may be to a full appreciation of the nature of the stimulus which induces a bird to set forth on its journey, that is, in a sense, only the beginning of the matter. The journey may be one of several thousand miles, perhaps in part across the sea, and in many instances accomplished by night.

Birds are charming creatures and they can become friendly, but not even their most fervent admirer will claim for them any great intelligence. And the future would certainly seem to belong to intelligence, if only because it permits a quicker assessment of any position and an immediate adaptation of behaviour to meet it. Adaptability is the key to success whenever the environment changes.

Now man is a social animal, and it will be interesting to compare his behaviour with that of a social insect such as the ant. Solomon urged the sluggard to " Go to the ant: consider his ways and be wise ". *What can the ant teach us?* Our answer is taken from Dr. Norbert Wiener's book on *The Human Use of Human Beings* in which he gave a popular account of his ideas on " Cybernetics ". Cybernetics, from the Greek word for a steersman, means the organization of communication and control, whether in a steam engine, a telegraph system, a calculating machine, or a human being. Dr. Wiener has suggested that the modern, high efficiency, electronic calculating machine, which can perform laborious computations far more quickly than a trained mathematician, might eventually throw some light on how the mind and the memory work—topics which keep puzzling physiologists and psychologists.

NORBERT WIENER

RIGIDITY AND LEARNING : ANTS AND MEN

WE have already said that the nature of social communities depends to a large extent upon their intrinsic modes of communication. The anthropologist knows very well that the patterns of communication of human communities are most various. There are communities like the Eskimos, among whom there seems to be no chieftainship and very little subordination, so that the basis of the social community is simply the common desire to survive on the part of individuals working against enormous odds of climate and food supply. There are socially stratified communities such as those of India, in which the relations of communication between two individuals are closely restricted and modified by their ancestry and position. There are the communities ruled by despots, in which every relation between two subjects becomes secondary to the relation between the subject and his king. There are the hierarchical feudal communities of lord and vassal, and the very special techniques of social communication which they involve.

In the ant community, each worker performs its proper functions. There is a separate caste of soldiers. Certain highly specialized individuals perform the functions of king and queen. If man were to adopt this community as a pattern, he would live in a Fascist state, in which ideally each individual is conditioned from birth for his proper occupation : in which rulers are perpetually rulers, soldiers perpetually soldiers, the peasant is never more than a peasant, and the worker is doomed to be a worker.

It is the thesis of this chapter that this aspiration of the Fascist for a human state based on the model of the ant is due to a profound misapprehension both of the nature of the ant and of the nature of man. I wish to show that the whole mode of development of the insect conditions it to be an essentially stupid and unlearning individual, cast in a mould which cannot be modified to any great extent ; and also how the physiological conditions of the ant make it into a cheap mass-produced article, of no more individual value than a paper pie plate to be thrown away after it is once used. On the other hand, I wish to show that the human individual re- presents an expensive investment of learning and study, extended under modern conditions for perhaps a quarter of a century, and almost half of his life. While it is possible to throw away this enormous advantage in training which the human being has, and which the ant does not have, and to organize the Fascist ant-state with human material, I shall indicate that this is a degradation of the very nature of man, and economically a waste of the greatest and most human values which man possesses.

I am afraid that I am convinced that a community of human beings is a far more useful thing than a community of ants ; and that if the human being is condemned and restricted to perform the functions of the ant and nothing more, he will not even be a good ant, not to mention a good human being. Those who would organize us according to permanent individual functions and permanent individual restrictions like those of the ant thereby condemn the human race to move at much less than half-steam. They throw away the greater part of our variability and of our chances for a reasonably long existence on this earth.

Let us now turn to a discussion of the facts and restrictions on the make-up of the ant which have made the ant community the very special thing it is. These facts and restrictions have a deep-seated origin in the anatomy and the physiology of the individual insect. Both the insect and the man are air-breathing forms, and represent the end of a long transition from the easy-going life of the water-borne animal to the much more exacting demands of the land. This transition from water to land, wherever it has occurred, has involved radical improvements in breathing, in the circulation generally, in the mechanical support of the organism, and in the sense organs.

The mechanical reinforcement of the bodies of land animals has taken place along several independent lines. In the case of most of the molluscs, as well as in the case of certain other groups which, though unrelated, have taken on generally a mollusc-like form, part of the outer surface secretes a non-living mass of calcareous tissue, the shell. This grows by accretion from an early stage in the animal until the end of its life. The spiral and helical forms of those groups need only this process of accretion to account for them.

If the shell is to remain an adequate protection for the animal, and the animal grows to any considerable size in its later stages, the shell must be a very appreciable burden, suitable only for land animals of the slowly moving and in-active life of the snail. In other shell-bearing animals, the shell is lighter and less of a load, but at the same time much less of a protection. The shell structure, with its heavy mechanical burden, has had only a limited success among land animals.

Man himself represents another direction of development —a direction found throughout the vertebrates, and at least indicated in invertebrates as highly developed as Limulus and the Octopus. In all these, certain internal parts of the connective tissue assume a consistency which is no longer fibrous, but rather that of a very hard, stiff jelly. These parts of the body are called *cartilage*, and they serve for the attach-ment of the powerful muscles which are needed for animals of an active life. In the higher vertebrates, this primary cartilaginous skeleton serves as a temporary scaffolding for a skeleton of much harder material : namely, bone, which is even more satisfactory for the attachment of powerful muscles. These skeletons, of bone or cartilage, contain a great deal of tissue which is not in any strict sense alive, but throughout this mass of intercellular tissue there is a living structure of cells, cellular membranes, and nutritive blood-vessels.

The vertebrates have developed not only internal skeletons, but also other features which suit them for active life. Their respiratory system, whether it takes the form of gills of the fish or the lungs of land-living vertebrates, is beautifully adapted for the active interchange of oxygen between the external medium and a blood, and the latter is made much more efficient than the average invertebrate blood by having its oxygen-carrying respiratory pigment concentrated in corpuscles. This blood is pumped through a closed system of vessels, rather than an open system of irregular sinuses, by a heart of relatively high efficiency. It is true that this situation is approached in certain invertebrates of high activity, such as the squids and the octopuses. It is most interesting that in other matters, such as the beginnings of a cartilaginous skeleton and the possession of effective image-forming eyes, these forms approximate the stage of organization found in the lower vertebrates.

The insects and crustaceans, and in fact all the arthropods, are built on quite another scheme of growth. The outer wall of the body is surrounded by a wall of chitin secreted by the cells of the epidermis. This chitin is a stiff substance rather closely related to cellulose. In the joints the layer of chitin is thin and moderately flexible, but over the rest of the animal it constitutes that hard external skeleton with which we are familiar in the case of the lobster and the cockroach. An internal skeleton can grow with the animal. An external skeleton (unless, like the shell of the snail, it grows by accretion) cannot. It is dead tissue, and possesses no intrinsic capability of growth. It serves to give a firm protection to the body and an attachment for the muscles. In other words, the arthropod lives within a strait-jacket.

Any internal growth can be converted into external growth only by discarding the old strait-jacket, and by developing under it a new one, which is initially soft and pliable and can take a slightly new and larger form, but which is very soon set to the rigidity of its predecessor. In other words, the stages of growth are marked by definite moults, relatively frequent in the crustacean, and much less so in the insect. There are several such stages possible during the larval period. The pupal period represents a transition moult, in which the wings which have not been functional in the larva develop internally towards a functional condition. This becomes realized when the pre-final pupal stage, and the moult which terminates it, give rise to a perfect adult. The adult never moults again. It is the sexual stage of the animal, and although in most cases it remains capable of taking nourishment, there are insects in which the adult mouth-parts and the digestive tube are aborted, so that the imago, as it is called, can only mate, lay eggs, and die.

In more primitive insects, such as the grasshopper or the cockroach, the larva is not essentially different in form from the imago ; the pupa is active, and the last moult differs from the others only in degree. In the higher or holometamorphic insects the difference between larva and adult is so profound that the transition involves a complete resting stage, and takes place catastrophically. The greater part of the tissues are attacked by leucocytes, the destroying white corpuscles of the blood. They resolve themselves into a structureless mush, which serves for the nutrition of a few parts called imaginal buds or discs. Structurally, these imaginal discs represent the whole of the imago. Thus the insect is reborn in a very real sense.

The nervous system takes part in this process of tearing down and building up. While there is a certain amount of evidence that some memory persists from the larva through to the imago, in the nature of the case this memory cannot be very extensive. The physiological condition for memory seems to be a certain continuity of organization, which allows the alterations produced by outer sense impressions to be retained as more or less permanent changes of structure or function. The erasure of metamorphosis is too radical to leave much permanent record of these changes. It is indeed hard to conceive of a memory of any precision which can survive this process of radical internal reconstruction.

There is another limitation on the insect, which is due to its method of respiration and circulation. The heart of the insect is a very poor and weak tubular structure, which opens, not into well-defined blood-vessels, but into vague cavities or sinuses conveying the blood to the tissues. This blood is without pigmented corpuscles, and carries the blood-pigments in solution. This mode of transferring oxygen seems to be definitely inferior to the corpuscular method.

In addition, the insect method of oxygenation of the tissues makes at most only local use of the blood. The body of the animal contains a system of branched tubules, carrying air directly from the outside into the tissues to be oxygenated. These tubules are stiffened against collapse by spiral fibres of chitin, and are thus passively open, but there is nowhere evidence of an active and effective system of air pumping. Respiration occurs by diffusion alone.

Notice that the same tubules carry by diffusion the good air in and the spent air polluted with carbon dioxide out to the surface. In a diffusion mechanism, the time of diffusion varies not as the length of the tube, but as the square of the length. Thus, in general, the efficiency of this system tends to fall off very rapidly with the size of the animal, and falls below the point of survival for an animal of any considerable size.

To follow the meaning of this limitation in size, let us compare two artificial structures—the cottage and the skyscraper. The ventilation of a cottage is quite adequately taken care of by the leak of air around the window frames, not to mention the draught of the chimney. No special ventilation is necessary. On the other hand, in a skyscraper with rooms within rooms, a shut-down of the system of forced ventilation will be followed in a very few minutes by an intolerable foulness of the air in the work spaces. Diffusion and even convection are no longer enough to ventilate such a structure.

In an entirely parallel way, the absolute maximum size of an insect is smaller than that attainable by a vertebrate. On the other hand, the ultimate elements of which the insect is composed are not always smaller than they are in man, or even in a whale. The nervous system partakes of this small size, and yet consists of neurons quite as large as those of the human brain. This is impossible unless there are many fewer neurons, and a much more limited complexity of structure. In the matter of intelligence, we should expect that it is not only the relative size of the nervous system which counts, but in a large measure its absolute size. There is simply no room in the reduced structure of an insect for a nervous system of great complexity, nor for a large stored memory.

In view of the impossibility of a large stored memory, as well as of the fact that the youth of an insect such as an ant is spent in a form which is insulated from the adult form by the intermediate catastrophe of metamorphosis, there is no opportunity for the ant to learn much. Add to this, that its behaviour in the adult stage must be substantially perfect from the beginning, and it then becomes clear that the instructions received by the insect nervous system must be substantially those due to the way it is built, and not to any personal experience. If compared with a computing machine, the insect must be the analogue of one with all its instructions set forth in advance on the " tapes ", and with a minimal opportunity for changing these instructions. This is meant when we say that the behaviour of an ant is much more a matter of instinct than of intelligence. In other words, the physical strait-jacket in which an insect grows up is directly responsible for the mental strait-jacket which regulates its pattern of behaviour.

Here the reader may say : " Well, we already know that the ant as an individual is not a very intelligent animal, so why all this fuss about explaining that it cannot be intelligent?" The answer is that the point of view of Cybernetics emphasizes the relation between the animal and the machine, and in the machine emphasizes the particular mode in which the machine functions as an index of what performance may be expected from it. Thus the fact that the mechanical conditions of performance of the insect are such as to limit the intelligence of the individual is highly relevant from the point of view of this book.

Even those naturalists like Fabre, to whom the emotional and the picturesque in the portrayal of the ant community have most appealed, have still kept their fundamental intellectual honesty and are forced to admit that the behaviour of the individual ant shows neither much originality nor much intelligence. A line of ants will play a game of follow the leader, but if there is no leader, and the line is made circular, they will continue to run around in this circle until they are exhausted. In general, among the insects, even when an insect of prey, such as the mantis, has already eaten off the abdomen of its victim, so that it is *in articulo mortis*, what is left of the animal will go on with its own task of eating as if nothing had happened. Not even the enthusiasm of the poet can remake the ant into a human individual.

In the matter of rigidity of behaviour, the greatest contrast to the ant is not merely the mammal in general, but man in particular. It has frequently been observed that man is a neoteinic form : that is, that if we compare man with the great apes, his closest relatives, we find that mature man in hair, head, shape, body proportions, bony structure, muscles, etc., is nearer to the new-born ape than to the adult ape.

Among the animals, man is a Peter Pan who never grows up.

This immaturity of anatomical structure corresponds to the longest relative period of childhood possessed by any animal. Physiologically, man does not reach puberty until he has already completed a fifth of his normal span of life. Let us compare this with the ratio in the case of a mouse, which lives three years and starts breeding at the end of three months. This is a ratio of twelve to one. The mouse's ratio is much more nearly typical of the large majority of mammals than is the human ratio.

Puberty for most mammals either represents the end of their period of tutelage, or is well beyond it. In our community, man is recognized as immature until the age of 21, and the modern period of education for the higher walks of life continues until about 30, actually beyond the period of most active physical strength. Man thus spends what may amount to 40 per cent. of his normal life as a learner. It is as completely natural for human society to be based on learning as for an ant society to be based on an inherited pattern. Learning is, in its essence, a form of feed-back, in which the pattern of behaviour is modified by past experience. Feed-back, as I have pointed out in the first chapter of my book, is a very general characteristic of forms of behaviour. In its simplest form, *the feed-back principle means that behaviour is scanned for its result, and that the success or failure of this result modifies future behaviour.* It is known to serve the function of rendering the behaviour of an individual or a machine relatively independent of the so-called " load " conditions.

Learning is a most complicated form of feed-back, and influences not merely the individual action, but the pattern of action. It is also a mode of rendering behaviour less at the mercy of the demands of the environment.

The great advantage which mankind possesses is thus the ability to learn. *What great achievement of the race has to be repeated by every baby?* He has to learn to speak. The question " What is man ? " with which this volume opened has been answered in many ways, but " the talking animal " is one of the best brief answers. Indeed, when the opponents of Darwin were looking around for arguments against his theory of the descent of man they made great use of the striking contrast between man's ability to express his thoughts in language and the dumbness of the apes. Here is an extract from a lecture delivered by Max Müller in 1861 in which this thought is developed. Müller was assuming, however, that man had evolved from some sub-human creature exactly like the great apes of to-day instead of from some species of *Homo* more ape-like than any living men, but also more man-like than any living apes. It is possible that the *Australopithecus* fossils, recently discovered in South Africa, may be some of the missing links in the chain.

MAX MÜLLER

THE IMPASSABLE BARRIER BETWEEN BRUTES AND MAN

IN comparing man with the other animals, we need not enter here into the physiological question whether the difference between the body of an ape and the body of a man is one of degree or of kind. However that question is settled by physiologists, we need not be afraid. If the structure of a mere worm is such as to fill the human mind with awe, if a single glimpse which we catch of the infinite wisdom displayed in the organs of the lowest creatures gives us an intimation of the wisdom of its Divine Creator far transcending the powers of our conception, how are we to criticize and disparage the most highly organized creatures of His creation, creatures as wonderfully made as we ourselves ?

Are there not many creatures on many points more perfect even than man ? Do we not envy the lion's strength, the eagle's eye, the wings of every bird ? If there existed animals as perfect as man in their physical structure, nay, even more perfect, no thoughtful man would ever be uneasy. His true superiority rests on different grounds.

> " I confess " Sydney Smith writes " I feel myself so much at ease about the superiority of mankind—I have such a marked and decided contempt for the understanding of every baboon I have ever seen—I feel so sure that the blue ape without a tail will never rival us in poetry, painting, and music, that I see no reason whatever that justice may not be done to the few fragments of soul and tatters of understanding which they may really possess."

The playfulness of Sydney Smith in handling serious and sacred subjects has been found fault with by many ; but humour is a safer sign of strong convictions and perfect sanity than guarded solemnity.

With regard to our own problem, no man can doubt that certain animals possess all the physical requirements for articulate speech. There is no letter of the alphabet which a parrot will not learn to pronounce. The fact, therefore, that the parrot is without a language of its own must be explained by a difference between the mental, not between the physical, faculties of the animal and man ; and it is by a comparison of the mental faculties alone, such as we find them in man and brutes, that we may hope to discover what constitutes the indispensable qualification for language, a qualification to be found in man alone, and in no other creature on earth.

I say mental faculties, and I mean to claim a large share of what we call our mental faculties for the higher animals. These animals have sensation, perception, memory, will, and intellect, only we must restrict intellect to the comparing or interlacing of single perceptions. All these points can be proved by irrefragable evidence, and that evidence has never, I believe, been summed up with greater lucidity and power than in one of the last publications of M. P. Flourens, *De la*

Raison, du Genie, et de la Folie (Paris, 1861). There are no doubt many people who are as much frightened at the idea that brutes have souls and are able to think, as by " the blue ape without a tail ". But their fright is entirely of their own making. If people will use such words as soul or thought without making it clear to themselves and others what they mean by them, these words will slip away under their feet, and the result must be painful. If we once ask the question, Have brutes a soul ? we shall never arrive at any conclusion ; for soul has been so many times defined by philosophers from Aristotle down to Hegel, that it means everything and nothing. Such has been the confusion caused by the promiscuous employment of the ill-defined terms of mental philosophy that we find Descartes representing brutes as living machines, whereas Leibnitz claims for them not only souls, but immortal souls. . . .

Instead of entering into these perplexities, which are chiefly due to the loose employment of ill-defined terms, let us simply look at the facts. Every unprejudiced observer will admit that—

(1) Brutes see, hear, taste, smell, and feel ; that is to say, they have five senses, just like ourselves, neither more nor less. They have both sensation and perception, a point which has been illustrated by M. Flourens by the most interesting experiments. If the roots of the optic nerve are removed, the retina in the eye of a bird ceases to be excitable, the iris is no longer movable ; the animal is blind, because it has lost the organ of sensation. If, on the contrary, the cerebral lobes are removed, the eye remains pure and sound, the retina excitable, the iris movable. The eye is preserved, yet the animal cannot see, because it has lost the organ of perception.

(2) Brutes have sensations of pleasure and pain. A dog that is beaten behaves exactly like a child that is chastised, and a dog that is fed and fondled exhibits

the same signs of satisfaction as a boy under the same circumstances. We can only judge from signs, and if they are to be trusted in the case of children, they must be trusted likewise in the case of brutes.

(3) Brutes do not forget, or, as philosophers would say, brutes have memory. They know their masters, they know their home; they evince joy on recognizing those who have been kind to them, and they bear malice for years to those by whom they have been insulted or ill-treated. Who does not recollect the dog Argos in the *Odyssey*, who, after so many years' absence, was the first to recognize Ulysses ?

(4) Brutes are able to compare and distinguish. A parrot will take up a nut, and throw it down again, without attempting to crack it. He has found that it is light : this he could discover only by comparing the weight of the good nuts with that of the bad ; and he has found that it has no kernel : this he could only discover by what philosophers would dignify with the grand title of syllogism, namely, " all light nuts are hollow, this is a light nut, therefore the nut is hollow ".

(5) Brutes have a will of their own. I appeal to anyone who has ever ridden a restive horse.

(6) Brutes show signs of shame and pride. Here again anyone who has to deal with dogs, who has watched a retriever with sparkling eyes placing a partridge at his master's feet, or a hound slinking away with his tail between his legs from the huntsman's call, will agree that these signs admit of but one interpretation. The difficulty begins when we use philosophical language, when we claim for brutes a moral sense, a conscience, a power of distinguishing good and evil ; and, as we gain nothing by these scholastic terms, it is better to avoid them altogether.

(7) Brutes show signs of love and hatred. There are well-authenticated stories of dogs following their masters

to the grave, and refusing food from anyone. Nor is there any doubt that brutes will watch their opportunity till they revenge themselves on those whom they dislike.

If, with all these facts before us, we deny that brutes have sensation, perception, memory, will, and intellect, we ought to bring forward powerful arguments for interpreting the signs which we observe in brutes so differently from those which we observe in man.

Some philosophers imagine they have explained everything if they ascribe to brutes instinct instead of intellect. But, if we take these two words in their usual acceptations, they surely do not exclude each other. There are instincts in man as well as in brutes. A child takes his mother's breast by instinct ; the spider weaves its net by instinct ; the bee builds her cell by instinct. No one would ascribe to the child a knowledge of physiology because it employs the exact muscles which are required for sucking ; nor shall we claim for the spider a knowledge of mechanics, or for the bee an acquaintance with geometry because we could not do what they do without a study of these sciences. But what if we tear a spider's web, and see the spider examining the mischief that is done, and either giving up his work in despair, or endeavouring to mend it as well as may be ? Surely here we have the instinct of weaving controlled by observation, by comparison, by reflection, by judgment. Instinct, whether mechanical or moral, is more prominent in brutes than in man ; but it exists in both, as much as intellect is shared by both.

Where, then, is the difference between brute and man ? What is it that man can do, and of which we find no signs, no rudiments in the whole brute world ? I answer without hesitation : the one great barrier between the brute and man is language. Man speaks, and no brute has ever uttered a word. Language is our Rubicon, and no brute will dare to cross it. This is our matter-of-fact answer to those who speak of development, who think they discover the rudiments at least of all human faculties in apes, and who would fain keep open the possibility that man is only a more favoured beast, the triumphant conqueror in the primeval struggle of life. Language is something more palpable than a fold of the brain, or an angle of the skull. It admits of no cavilling, and no process of natural selection will ever distil significant words out of the notes of birds or the cries of beasts.

Language, however, is the only outward sign. We may point to it in our arguments, we may challenge our opponent to produce anything approaching to it from the whole brute world. But if this were all, if the art of employing articulate sounds for the purpose of communicating impressions were the only thing by which we could assert our superiority over the brute creation, we might not unreasonably feel somewhat uneasy at having the gorilla so close on our heels.

It cannot be denied that brutes, though they do not use articulate sounds for that purpose, have, nevertheless, means of their own for communicating with each other. When a whale is struck, the whole shoal, though widely dispersed, are instantly made aware of the presence of an enemy ; and when the gravedigger beetle finds the carcass of a mole, he hastens to communicate the discovery to his fellows, and soon returns with his four confederates. It is evident, too, that dogs, though they do not speak, possess the power of understanding much that is said to them—their names and the calls of their masters ; and other animals, such as the parrot, can pronounce every articulate sound. Hence, although for the purpose of philosophical warfare, articulate language would still form an impregnable position, yet it is but natural that for our own satisfaction we should try to find out in what the strength of our position really consists ; or, in other words, that we should try to discover that inward power of which language is the outward sign and manifestation.

For this purpose it will be best to examine the opinions of those who approached our problem from another point ; who, instead of looking for outward and palpable signs of difference between brute and man, inquired into the inward mental faculties, and tried to determine the point where man transcends the barriers of the brute intellect. That point, if truly determined, ought to coincide with the starting-point of language ; and, if so, that coincidence ought to explain the problem which occupies us at present.

I shall read an extract from Locke's *Essay Concerning Human Understanding.*

After having explained how universal ideas are made, how the mind, having observed the same colour in chalk, and snow, and milk, comprehends these single perceptions under the general conception of whiteness, Locke continues :

> " If it may be doubted whether beasts compound and enlarge their ideas that way to any degree, this, I think, I may be positive in, that the power of abstracting is not at all in them ; and that the having of general ideas is that which puts a perfect distinction betwixt man and brutes, and is an excellency which the faculties of brutes do by no means attain to."

If Locke is right in considering the having general ideas as the distinguishing feature between man and brutes, and, if we ourselves are right in pointing to language as the one palpable distinction between the two, it would seem to follow that language is the outward sign and realization of that inward faculty which is called the faculty of abstraction, but which is better known to us by the homely name of reason.

Müller was right in suggesting that language is but the outward sign of the faculty of reason. But other evidence of the reasoning powers of primitive man is available in his manufacture of tools. Indeed, Dr. Kenneth Oakley, the author of the British Museum handbook, *Man the Tool-Maker,* seems to think that the ability to make (not just to use) tools is probably the best sign differentiating man from his more primitive and sub-human ancestors. We must not expect to find any clear-cut line of division. The precursors came down from the trees and stood upright, adapting themselves to life in open country. Thus the hands were set free to make use of natural tools—stones, or branches of trees, or the long bones of animals. A change in diet also occurred, with flesh-eating become common, probably during severe drought, and with this came the need for weapons to kill the animals and for edged tools to cut up the carcase. When such tools and weapons were made in advance, with a clear realization of how and when they would be used, *Homo sapiens* had arrived. Verbal language probably came later. *Why might Man be described as " the tool-maker " ?*

KENNETH P. OAKLEY

A DEFINITION OF MAN

THE definition of man as the tool-making primate is sometimes criticized on the score that (1) lower primates occasionally make use of tools, (2) tool-making is merely one of the by-products of man's mental development, and not a primary characteristic from a zoological point of view. It is worth while to examine these criticisms in some detail.

The *use* of tools and weapons is certainly not confined to man. There are records of monkeys throwing sticks and stones ; a baboon will sometimes crack open a scorpion with a pebble ; while some of the apes observed by Köhler used sticks as levers for digging up objects hidden in the ground, and for extending their reach. The use of tools is not even confined to primates : when the southern sea-otter finds on the sea-floor a certain hard-shelled mollusc, it takes it to the surface, together with a stone to serve as an anvil on which to crack it open.

The *manufacture* of tools requires mental activity of a different order. The chimpanzee is the only ape reliably reported to make tools, and then only in captivity. Sultan, one of the male chimpanzees observed by Köhler, fitted a small bamboo cane into a larger one so as to make a stick long enough to secure a bunch of bananas which could not be reached by the use of either of the rods used singly. Once he attained the same result by fitting into one of the canes a piece of wood which he pointed for the purpose with the aid of his teeth. Apes are thus evidently capable of improvising tools. But it is important to note that all the improvisations effected by Sultan were carried out with *visible* reward as incentive. Köhler could obtain no indication that apes are ever capable of conceiving the usefulness of shaping an object for use in an imagined future eventuality. He expressed his conclusions as follows :

> " The time in which the chimpanzee lives is limited in past and future . . . . it is in the extremely narrow limits in *this* direction that the chief difference is to be found between anthropoids and the most primitive human beings. The lack of an invaluable technical aid (speech) and a great limitation of those very important components of thought, so-called ' images ', would thus constitute the causes that prevent the chimpanzee from attaining even the smallest beginnings of cultural development."

In other words, apes are very limited in their capacity for visualizing and thinking about the relationships between objects when those objects are not in sight. The power of abstraction or conceptual thought is basic to the regular manufacture of tools. In apes it is no more than nascent. Madame Kohts in Moscow found that her chimpanzee could select from a collection of objects of extremely varied form those which were of the same shade of colour. This mental isolation of a single feature from a varied field of observation is the dawn of conceptual thought ; but Mme Kohts found that ideas such as this rapidly fade in the mind of the ape.

To illustrate more clearly the difference between the mental capacity of an ape and of a human being, the following observations of Köhler are worth quoting. When embarrassed by lack of a stick, a chimpanzee will pull a loose board from an old box and use it. If the boards are nailed together in such a way that they make an unbroken surface, the chimpanzee, although strong enough to break up the box, will not see " possible sticks " in it—even if his need of a stick is urgent. Men, on the other hand, seeking to make a tool of a form suited to a particular purpose will visualize its shape in a form-less lump of stone and chip it until the imagined tool is actualized.   There is the possibility of gradation between these two extremes, perceptual thought in apes, conceptual thought in man ; but it seems necessary to stress the contrast because one is apt to be so impressed by the occasional manufacture of tools by apes that there is a danger of mini-mizing the gap in the quality of mind needed for such efforts, compared with even the crudest tools of early man, which indicate forethought.

Language must have greatly facilitated the development of systematic manufacture of tools.   Oral tradition, in effect a new kind of inheritance, is sometimes regarded as more distinctive of man than tool-making.   But speech itself is really a class of tool, " an invaluable technical aid ", in Köhler's description, and depends on a capacity for conceptual thought.   The brains of the earliest tool-making Hominidæ were probably advanced enough functionally for speech, but nevertheless verbal language may have been a comparatively late cultural development—an invention.   The earliest means of com-municating ideas was perhaps by gesticulation, mainly of mouth and hands, accompanied by cries and grunts to attract attention (" gesture language ").

On the evidence available it appears reasonable to conclude that systematic tool-making, in the broadest sense, would only be possible in primates whose level of cerebral development was in advance of that of modern apes.

It remains to consider at what point in the evolution of the Hominidæ did tool-making arise, and whether it can be classed as a fundamental characteristic of man from a zoological point of view. It might be thought that manual dexterity was a limiting factor. Actually, however, the prehensile hands of the less specialized monkeys would be quite capable of most of the ordinary movements in making and using tools, if directed by an adequate brain.

Our unspecialized, five-fingered hands are so well adapted to grasping that they are a clear indication that our remoter ancestors were accustomed to climbing in trees. So long as they continued to lead an arboreal life their prehensile hands, fully occupied by climbing and feeding, had no opportunity or need to develop the habit of using external objects as functional extensions of the limbs. However, some of our early ancestors must have become adapted to spending part of their time walking and sitting on open ground. Their hands thus became free to handle objects, first out of idle curiosity, later perhaps to some purpose. Some of the Miocene apes were agile monkey-like creatures accustomed to running on the ground as well as climbing trees, and probably capable, too, of rearing up on hind legs. Such apes may well have been in the habit of improvising tools when circumstances demanded. It is suggestive that the chimpanzee, not known to be a tool-user in its forest environment, makes use of sticks for various purposes as soon as captivity forces it to spend most of its time on the ground.

From a functional point of view tools may be regarded as detachable extensions of the forelimb. Most mammals have evolved specialized bodily equipment suited to some particular mode of life. Horses, for example, have teeth suited to eating grass, and hoofs adapted for galloping over hard plains and thus avoiding attack. The carnivorous sabre-tooth cats evolved claws like grappling-irons and canine teeth like daggers, perfectly adapted for killing prey. In process of evolution the Hominidæ avoided any such extreme specialization, their teeth being suited to an omnivorous diet, while they retained the pliant five-fingered hands which were so useful to their small tree-dwelling ancestors. When the forerunners of man acquired the ability to walk upright habitually, their hands became free to use tools.

One can imagine that " proto-hominids " may have remained at the stage of occasional tool-using for millions of years. Systematic tool-making would not have followed until the brain had attained the complexity of organization requisite for conceptual thought ; and even then, this activity would be unlikely to become regular until new habits were demanded by circumstances.

Probably within the Pliocene period, certainly by early Pleistocene times, the human level of cerebral development had been reached, and stone artifacts of standardized types were being made. These indicate that the manufacture of tools served certain permanent needs of the earliest human beings. What were these needs ? The use of tools and weapons was surely the means whereby the Hominidæ kept themselves alive after they abandoned the protection and sustenance provided by forests.

Tree-climbing primates had no use for tools. Tool-using arose in connexion with adaptation to life on open ground away from forests. In the evolution of the primates the forelimbs have continually shown a tendency to take on functions performed in their ancestors by the teeth. The use of tools is evidently an extension of this trend, and we may suppose that tools were largely substitutes for teeth.

The apes of to-day are forest creatures, subsisting almost exclusively on fruits, leaves, shoots and insects. All known races of man, on the other hand, include a substantial proportion of animal flesh in their diet. We have ample evidence that Pekin Man, Neanderthal Man, and Late Palæolithic races of *Homo sapiens* were meat-eaters. The discoveries of broken meat-bones on the Early Pelæolithic camping sites at Olorgesaile in Kenya have shown that the Acheulian handaxe-makers of these sites (early precursors of *Homo sapiens*) were also hunters. I suggest that meat-eating is as old as man ; that, with the change from forest living to open country, the diet of proto-men inevitably became more varied and that they changed from being eaters largely of plants and the fruits of plants to being partly meat-eaters.

It seems probable, on the analogy of the baboons, that early Hominidæ living in country like the present African savannah may have become addicted to flesh-eating as a result of the struggle for existence being intensified by excessive drought. It may be recalled that baboons, almost the only monkeys completely adapted to life away from woodlands, occasionally prey on lambs and other animals of similar size, using their powerful canine teeth as offensive weapons, and moreover that this habit is liable to become more prevalent when conditions of existence are hard. Owing to the extensive folk-lore associated with baboons, reports of the carnivorous habits of those in South Africa have been discounted by some zoologists, but information from many different observers, recently collected by my friend Mr. F. E. Hiley, leaves no doubt that such reports are substantially true. A report from Captain H. B. Potte, Game Conservator in Zululand, is typical of those received :

" The following are my personal observations over a period of twenty years' wardenship in the Hluhluwe Game Reserve : I have seen full-grown poultry killed and actually eaten by baboons, mostly however by aged individuals. Eggs and chickens are taken by the dozen, by old and young baboons. I have on many occasions actually witnessed apparently organized hunts which often result in the death of the intended victim. The baboons, usually led by a veteran of the troop, surround an unsuspecting three-parts grown Mountain Reedbuck, or Duiker, as the case may be, and on one occasion a young Reedbuck doe was the victim. It would appear that on a given signal the baboons close in on their quarry, catch it and tear it asunder. As a matter of interest I have refrained from interference in these grim encounters so that I would be in a position authentically to record the results. In nine cases out of ten the game animal is devoured limb by limb and after the affair is over all that is to be found are the skull and leg bones."

Baboons, like some other monkeys and apes too, have powerful canine teeth, serving mainly for defence against carnivores and for attack in mating duels. It has been suggested that the reduction of the canine teeth in man was an outcome of the use of hand-weapons. But judging by their condition in the australopithecines, the canine teeth were reduced in the Hominidæ at an evolutionary stage below that of tool-making. Certainly the " proto-hominids " would have needed some means of defending themselves in the open, and having their hands free may well have used stones as missiles and sticks or animal long-bones as clubs.

In dry open country (such, for instance, as that inhabited by the australopithecines) the " proto-hominids ", like baboons, might readily have taken to eating flesh, particularly in times of drought. Although they lacked teeth suited to carnivorous habits they could easily have killed small mammals. Life in the open set a premium on co-operation. Drawing on our knowledge of the mentality and social life of other primates, it seems not unreasonable to suppose that, hunting in hordes, the " proto-hominids " could have killed medium-sized mammals, say by cornering them and using improvised hand-weapons such as they might earlier have learnt to improvise for their own defence. This is, frankly, speculation. Is there any real evidence that the early Hominidæ were carnivorous and that they passed through a tool-using stage, before becoming tool-makers?

Even if the australopithecines are geologically too late to be actual ancestors of Man, their " morphological dating " is plain—structurally, they may be viewed as slightly modified descendants of our Pliocene forebears. There are indications that they may have been carnivorous, and that possibly they used improvised tools and weapons. The quantities of broken animal bones and fragments of egg shells found with *Australopithecus* in the cave deposits at Taungs suggest a " midden " of food refuse; but the possibility cannot be ruled out that the bones were introduced by a carnivore which left no other trace of its presence. Professor Dart claims that some of the long-bones of antelopes found in the australopithecine layer at Makapan show signs of having been used as weapons. But again, although there is a strong probability that these small bipedal primates were tool-users, the evidence remains *sub judice*.

By the time that the Hominidæ had evolved into toolmakers they were evidently largely carnivorous—quantities of meat-bones were associated with the remains of *Pithecanthropus pekinensis*. It is easy to see how the one habit led from the other. Although the killing of game may have been accomplished easily enough in some such way as that suggested above, the early hominids must often have encountered difficulty in removing skin and fur, and in dividing the flesh. In the absence of strong canine teeth the solution would have been overcome most readily by using sharp pieces of stone. Here surely was the origin of the tradition of tool-making. Where no naturally sharp pieces of stone lay ready to hand, some of the more intelligent individuals saw that the solution was to break pebbles and produce fresh sharp edges. Once the tradition of tool-making had begun, the manifold uses of chipped stones became obvious ; they were useful for sharpening sticks for use in digging out burrowing mammals, for making spears sharp enough to be effective weapons in hunting larger game, for scraping meat from bones, splitting them to get at the marrow, chopping the meat into convenient mouthfuls. All the main uses of stone tools were, I suggest, connected in the first place with adoption of semi-carnivorous habits.

From the endowment of nature we should be vegetarians. We lack the teeth evolved by true carnivores, and we have the long gut associated with a herbivorous diet. Furthermore, we are the only members of the Hominoidea which are accustomed to eat meat on any considerable scale. It is true that anthropoid apes, like most herbivores, consume small quantities of animal protein ; some of them occasionally rob birds' nests of eggs and fledglings, but by and large they are fruit and plant eaters.

One can well imagine that a changing environment, for instance during a period of desiccation, may have produced an abnormal appetite in the early hominids. Gorillas in captivity quickly develop a liking for meat, and this appears to be due to a change in the flora and fauna of their intestines. Normally their intestines are richly supplied with ciliate protozoa (Infusoria), which serve to digest cellulose. According to Reichenow, under the abnormal conditions of captivity the Infusoria are ingested, and with their disappearance from the intestine the animal develops an abnormal appetite, and readily takes to eating meat—and may even prefer meat to its normal fare.

By widening his diet and becoming a tool-maker man became the most adaptable of all primates. The change from herbivorous to semi-carnivorous habits was important from the point of view of the use of energy. To obtain a given amount of energy a carnivore subsists on a smaller quantity of food than a herbivore. Instead of eating almost continuously like their ancestors, the Hominidæ spent much of their time in hunting. This led to increased interdependence. New skills and aptitudes were developed through this new way of life, and with increasing control of environment through the use of tools, man became the most adaptable of all creatures, and free to spread into every climatic zone.

If man is a thinking, talking, and tool-making animal, we might reasonably ask, as an interesting sidelight on this aspect of his powers, if he can communicate with his fellows in any other way. *Is there anything in telepathy?* Many experiments made in recent years seem to show that certain people do have the power to read what is in another person's mind. It has indeed been suggested that some such method of communicating vague ideas was used before the power of speech was evolved, and that it has gradually faded through disuse as the more exact method developed. On the other hand, it may be something new which is only just appearing as man becomes more " sensitive " to the feelings of his fellow-men. Many of our readers will have had experiences which can be most easily explained by supposing that some strange communication quite outside the everyday channels did occur ; and the mystery remains even if we follow the unromantic custom of this sophisticated age and refer to the phenomena by initials : E.S.P., short for extra-sensory perception.

An answer to this question is provided by Dr. S. G. Soal, the recipient of the first D.Sc. degree ever given by a British university for psychical research. His lecture also suggests that the power to communicate by telepathy may also be allied to the power to foresee future events—" precognition " as it is called by the para-psychologists, while the sceptics murmur something about " remarkably good guesses ".

SAMUEL GEORGE SOAL

SEEING INTO FUTURE TIME

IN the year 1934 Dr. Rhine of Duke University, North Carolina, dropped a bomb among the psychologists. He claimed to have demonstrated telepathy by making people guess at the geometrical diagrams on cards. Well, I suppose most people have come across cases of telepathy either at first-hand or by hearsay. Let me tell you an example. A child of ten was walking along a country lane reading a book on geometry when suddenly her surroundings faded away and she saw her mother lying on the floor of a disused room at home apparently dead with a lace handkerchief beside her. The child was so impressed that, instead of returning straight home, she rushed to the doctor's and brought him back with her. They found the woman lying on the floor of the room mentioned suffering from a severe heart attack. The doctor was just in time to save her life. Beside her was the lace handkerchief.

Such a case as this is enormously interesting, but there is not much the scientist can do about it. It is always possible that the child might have invented part of the story after the fact, but even if the story is strictly accurate the scientist cannot repeat the experience and vary the conditions under which it happened. And this is the essence of scientific experiment. We might go on collecting cases like the little girl's for a couple of centuries and still be no nearer understanding them. If we want to know anything about telepathy we must not be content merely to observe ; we must do experiments. When Dr. Rhine published his findings I was extremely sceptical. I knew the snags in this sort of experiment. I had been testing telepathy by mathematical methods ever since 1928 and always without success. I wondered, among other things, whether Rhine had taken sufficient care to prevent his guessers from recognizing the cards by means of, say, specks or irregularities on their backs. But I decided to repeat the experiments and to use every possible safeguard.

Now in a card-guessing experiment two persons are essential. You must have a person who looks at the cards— known as the " agent "—and another person called the " guesser " who tries to read the agent's thoughts. I seated my agent on one side of an opaque screen while the guesser or thought-reader sat on the other side. A playing-cards firm had made me cards whose faces bore the five symbols : cross, circle, oblong, star, and wavy lines. There were two hundred and forty cards of each symbol—twelve hundred altogether. Now, when the experiment began, the twelve hundred cards were arranged haphazard and gathered into packs of twenty-five. I took a pack and laid it face downwards on the table. Closing my eyes, I lifted off the cards one by one, exposing each card for a few seconds to the gaze of the agent. At a

signal from me, the man on the other side of the screen had to
guess which of the five symbols the agent was looking at.
By sheer chance alone, without any help from telepathy, he
might expect to score, on average, five correct hits in twenty-
five guesses. But if he maintained an average of, say, seven or
eight correct hits over a considerable number of packs of
twenty-five, that would prove something more than luck was
helping him. Well, I tested a hundred and sixty people and
collected over a hundred and twenty-eight thousand guesses.
I stuck to the job for five solid years testing psychic Indians,
inscrutable Chinese, Jews and Moslems, vaudeville tele-
pathists, spiritualistic mediums, mystic Egyptians, hypno-
tizable Celts, people who had been blind from birth and people
suffering from nervous breakdown, and *not a trace* of telepathy
among the whole lot ! After five years I felt I deserved a rest.
I had done my share.

I was left alone for three months. Then came along
that super-enthusiast, Whately Carington. He wanted me to
see if any of the people had been guessing correctly—not the
card the agent was looking at, but perhaps the card he just
had been looking at, or the one he was just about to look at—
the card one ahead or one behind. It was as if a rifleman
might possess a bias which caused his shots to strike the target
persistently to either the right or the left of the bull's-eye.
Well, Carington kept on urging me to re-examine my data.
I started on this job feeling hopelessly bored. But I soon
stopped being bored ; I became interested and then aston-
ished, because I discovered that two of my hundred and sixty
guessers had actually been guessing the card one ahead and
the card one behind, and doing this to an extent which excluded
chance as an explanation. One of these two was Basil Shackleton, the West End photographer.

Now the fact that Shackleton had been guessing correctly
so many of the cards one place ahead did not prove that he
was seeing into the future, for the card was lying face downwards on top of the pack at the instant when he guessed it

correctly, and it was possible that he was getting his knowledge from the card itself by clairvoyance. But he might have been foreseeing the mental picture which would be in the agent's mind in two or three seconds' time. We had to decide between these two possibilities. To test the matter, Mrs. K. M. Goldney, a friend of mine, and I embarked on a whole series of fresh experiments. Let me describe one of them. You will remember what I said about the necessity of safeguards. The lay-out I am going to describe may sound rather elaborate, but it was indispensable. I had to close every loophole against trickery and leakage of information. I had to make quite certain that the man who looked at the cards was giving nothing away to the guesser, either intentionally or by accident.

Imagine two rooms—a large one called the studio and a smaller one called the ante-room—separated by one dividing wall with a door in it. Mr. Shackleton sits in the ante-room with a sheet of paper before him on which he writes down his guesses. Picture him as a tallish, dark man whose every movement displays the nervous irritability of the artist. In the centre of the studio a vertical screen stands on a small table. In the centre of the screen is cut a small square hole. On the side of the screen farthest from the dividing wall sits the agent in front of five cards bearing pictures of the five animals : lion, elephant, giraffe, zebra, pelican. To prevent the cards being seen by anyone but the agent they are not only face-downwards on the table but they are covered by a box whose only open side faces the agent. On the near side of the screen stands Experimenter No. 1 with a bowl which contains a large number of coloured counters in five different colours, equal numbers of each colour. In the ante-room, Experimenter No. 2 sits beside Shackleton.

Unseen by anyone else, the agent shuffles the five cards and lays them face downwards in their row in the box. It is agreed that the five colours white to blue are to stand for the positions one to five of the cards in the box counting from left to right. Looking straight ahead, Experimenter No. 1 takes a counter from the bowl, presents it at the hole in the screen, pauses a fraction of a second and calls in a loud voice, " First guess ". When the agent sees the counter at the hole—say a red one—he raises slightly the fourth card in the row, looks at its face and lets it fall face downwards into its place. Red stands for four. When he hears " First guess," Shackleton immediately writes down in the first square on his paper one of the five letters L, E, G, Z, or P, these being the initials of the five animal names. Experimenter No. 1 drops the counter into the bowl, draws out another, shows it at the hole and calls, " Number two ". The agent looks at the appropriate card and Shackleton writes down another initial letter in the second space of his paper. As each counter is presented at the hole, a person sitting where he cannot see the cards records the colour of the counter.

Experimenter No. 1 goes on drawing out counters at intervals of about two and a half seconds, and when fifty guesses have been made he goes round the screen, turns up the five cards and records their order : say, L, G, P, E, Z. The agent then reshuffles the cards out of everyone's sight and another sheet of fifty guesses is finished in the same way. After about two hundred guesses the experimenters and witnesses come together and compare Shackleton's list of guesses with the actual order of animal pictures seen by the agent. A count is made of the number of guesses correct on the actual card, and the card one ahead and the card one behind.

We found that, working with our first agent, Miss Rita Elliott, Shackleton got the card one ahead correct on an average between eight and nine times in twenty-five trials. After a few weeks' work the odds against chance had mounted to astronomical figures. (Remember that Experimenter No. 1 who gave the signal when to guess could give Shackleton no help by the inflections of her voice because she did not know the order of the cards in the box.) But we soon satisfied ourselves that, for the experiment to succeed, it was essential for the agent to know the order of the cards in the box. If the agent did not know their order and was not allowed to look at the cards as the counters were shown, the experiment failed. This suggested that clairvoyance was not the explanation of success. But even when the agent looked at the pictures the experiment did not succeed with everyone acting as agent. Some people were very successful as transmitters or agents ; others were no use at all. All this indicated that Shackleton obtained his knowledge of the card one ahead by foreseeing the mental picture which would be in the agent's mind in about two and a half seconds' time.

One day we made an astonishing discovery. We tried the effect of speeding up the presentation of the counters so that the interval between successive counters was reduced from two and a half seconds to about one and a half seconds, and— still with Miss Elliott as agent—we found that Shackleton ceased to score on the next card but began to guess correctly the next but one ! Now at the instant Shackleton made his guess the counter whose colour determined the card to be looked at two places ahead could not possibly have been selected by the experimenter : Shackleton's guess was being decided by a mental event in the agent's mind which did not take place till three seconds later. Apparently the law of causation was being reversed—a result was preceding in time a cause of which it was to be the effect. The shift of the precognitive effect from a plus 1 card to a plus 2 card noted with Miss Elliott was fully confirmed when Mr. J. Aldred acted as agent.

I do not think these experiments are open to any attack. The mathematical theory of probability has been in use for a great many years in checking the results of card-guessing experiments ; in fact card-guessing is an ideal field for the exercise of the mathematical theory. Any critic who tried to demolish our work on mathematical grounds would first find it necessary to overthrow the accepted statistical practice of the last fifty years—and this would give him a severe headache. More than twenty intelligent observers took part in the experiments, including philosophers like Professor Habberley Price of Oxford and psychologists like Professor Mace of London ; but no one could find a flaw. It seems to me that our experiments now make it almost hopeless to attempt to explain telepathy by means of electro-magnetic waves or radiations emanating from one brain and impinging upon another. In our experiments the " message " was received by Shackleton and acted upon before it could have been conceived, let alone transmitted by the agent.

I have no doubt that the materialists will maintain that such things cannot happen, since they upset their idea of physical causation. Of course they do, and that is just why they are to me so stimulating and important. There is no reason whatever to suppose that the material world embraces the whole of reality. Our experiments are to me pointers to a universe of reality which lies outside the physical order. Form the mental picture of a house you lived in maybe thirty years ago. This mental picture is real, but it is not inside your head or outside it. It is not in physical space at all. Or again, who could give a physical description of the meaning of a Beethoven sonata ? There are non-physical realities. I believe that, in this extra-physical world of mental images in which telepathy operates, our human personality has its deepest roots, and in this world it may still survive after the physical organism has ceased to function.

BOOK III
BIOLOGY AND HEALTH

BIOLOGY AND HEALTH

HOWEVER colourful the thread spun by Lachesis, the time always comes, sooner or later, for Atropos to use her shears. And in former ages the fatal moment came very early. Every baby born in Britain, Canada, or Australasia to-day can expect to live some sixty years or more—a dozen years more than its parents and fifteen years more than its grandparents. A century ago the expectation of life was about 40 years ; in the eighteenth century it was no more than 26 years, and in the Middle Ages 21. Disease, famine, warfare, and the accidents of life cut short many a promising existence.

Now the biologist, as such, cannot prevent warfare or accident, but he can deal with disease, and he has done much to prevent famine. And in this last book we shall be mainly concerned with the ways in which he has beguiled Atropos into postponing her malicious, but inevitable, snip.

First let us deal with disease. *How were diseases regarded in pre-scientific days ?* For an answer to this question we cannot do better than consult Robert Burton, the witty and satirical author of *The Anatomy of Melancholy*. This book, first published in 1621, gives a view on diseases which would have been accepted by most men not only of his own day and of a century or two later, but of two or three thousand years earlier. And mixed with all the superstitions is not a little sound common sense.

ROBERT BURTON

DISEASES IN GENERAL

THE instrumental causes of these our infirmities are as divers as the infirmities themselves. Stars, heavens, elements, etc., and all those creatures which God hath made, are armed against sinners. They were indeed once good in themselves, and that they are now many of them pernicious unto us, is not in their nature, but our corruption, which hath caused it. For, from the fall of our first parent Adam, they have been changed, the earth accursed, the influence of stars altered, the four elements, beasts, birds, plants, are now ready to offend us. " The principal things for the use of man are water, fire, iron, salt, meal, wheat, honey, milk, oil, wine, clothing, good to the godly, to the sinners turned to evil." " Fire, and hail, and famine, and dearth, all these are created for vengeance." The heavens threaten us with their comets, stars, planets, with their great conjunctions, eclipses, oppositions, quartiles, and such unfriendly aspects ; the air with his meteors, thunder and lightning, intemperate heat and cold, mighty winds, tempests, unseasonable weather ; from which proceed dearth, famine, plague, and all sorts of epidemical diseases, consuming infinite myriads of men.

At Cairo in Egypt, every third year (as it is related by Boterus, and others) 300,000 die of the plague ; and 200,000 in Constantinople, every fifth or seventh at the utmost. How doth the earth terrify and oppress us with terrible earthquakes, which are most frequent in China, Japan, and those Eastern Climes, swallowing up sometimes six cities at once ! How doth the water rage with his inundations, irruptions, flinging down towns, cities, villages, bridges, etc., besides shipwrecks ; whole islands are sometimes suddenly overwhelmed with all their inhabitants in Zealand, Holland, and many parts of the Continent drowned, as the Lake Erne in Ireland ! " And we perceive nothing except the remains of cities in the open sea." In the fens of Friesland, 1230, by reason of tempests, the sea drowned all the country almost, men and cattle in it. How doth the fire rage, that merciless element, consuming in an instant whole cities ! What town, of any antiquity or note, hath not been once, again and again, by the fury of this merciless element, defaced, ruinated, and left desolate ? In a word,

> " Whom fire spares, sea doth drown ; whom sea,
> Pestilent air doth send to clay ;
> Who war scapes, sickness takes away."

To descend to more particulars, how many creatures are at
deadly feud with men ! Lions, wolves, bears, etc., some with
hoofs, horns, tusks, teeth, nails. How many noxious serpents
and venomous creatures, ready to offend us with stings, breath,
sight or quite kill us ! How many pernicious fishes, plants,
gums, fruits, seeds, flowers, etc., could I reckon up on a
sudden, which by their very smell, many of them, touch,
taste, cause some grievous malady, if not death itself ! Some
make mention of a thousand several poisons : but these are
but trifles in respect. The greatest enemy to man is man,
who by the Devil's instigation is still ready to do mischief,
his own executioner, a wolf, a Devil to himself and others.
We are all brethren in Christ, or at least should be, members
of one body, servants of one Lord, and yet no fiend can so
torment, insult over, tyrannize, vex, as one man doth another.
Let me not fall therefore (saith David, when wars, plague,
famine were offered) into the hands of men, merciless and
wicked men :

" Scarce are they worthy of the name of men,
For fiercer far are they than ravening wolves."

We can most part foresee these epidemical diseases, and
likely avoid them. Dearths, tempests, plagues, our Astrologers
foretell us ; earthquakes, inundations, ruins of houses,
consuming fire, come by little and little, or make some noise
before-hand ; but the knaveries, impostures, injuries, and
villainies, of men no art can avoid. We can keep our professed
enemies from our cities, by gates, walls, and towers, defend
ourselves from thieves and robbers by watchfulness and
weapons ; but this malice of men, and their pernicious
endeavours, no caution can divert, no vigilancy foresee, we
have so many secret plots and devices to mischief one another.
Sometimes by the Devil's help, as Magicians, Witches :
sometimes by impostures, mixtures, poisons, stratagems,
single combats, wars, we hack and hew, as if we were like
Cadmus' soldiers, born to consume one another. 'Tis an

ordinary thing to read of a hundred and two hundred thousand men slain in a battle ; besides all manner of tortures, brazen bulls, racks, wheels, strappadoes, guns, engines, etc. We have invented more torturing instruments than there be several members in a man's body, as Cyprian well observes. To come nearer yet, our own parents by their offences, indiscretion, and intemperance, are our mortal enemies. The fathers have eaten sour grapes, and the children's teeth are set on edge. They cause our grief many times, and put upon us hereditary diseases, inevitable infirmities : they torment us, and we are ready to injure our posterity ;

" Like to produce still more degenerate stock ",

and the latter end of the world, as Paul foretold, is still like to be worst. We are thus bad by nature, bad by kind, but far worse by art, every man the greatest enemy unto himself. We study many times to undo ourselves, abusing those good things which God hath bestowed upon us, health, wealth, strength, wit, learning, art, memory, to our own destruction. As Judas Maccabæus killed Apollonius with his own weapons, we arm ourselves to our own overthrows ; and use reason, art, judgment, all that should help us, as so many instruments to undo us.

Hector gave Ajax a sword, which, so long as he fought against enemies, served for his help and defence ; but after he began to hurt harmless creatures with it, turned to his own hurtless bowels. Those excellent means God hath bestowed on us, well employed, cannot but much avail us ; but if otherwise perverted, they ruin and confound us : and so by reason of our indiscretion and weakness they commonly do, we have too many instances.

If you will particularly know how, and by what means, consult Physicians, and they will tell you, that it is in offending in some of those six non-natural things, of which I shall after dilate more at large ; they are the causes of our infirmities, our surfeiting, and drunkenness, our immoderate insatiable

lust, and prodigious riot. " Plures crapula quam gladius ",
is a true saying, the board consumes more than the sword.
Our intemperance it is that pulls so many several incurable
diseases upon our heads, that hastens old age, perverts our
temperature, and brings upon us sudden death. And last
of all, that which crucifies us most, is our own folly, madness,
(" whom Jupiter desires to destroy, he first drives mad "),
weakness, want of government, our facility and proneness in
yielding to several lusts, in giving way to every passion and
perturbation of the mind : by which means we metamorphose
ourselves, and degenerate into beasts.

For Burton an illness was something which came by Divine command, and there was nothing to be done to prevent it. The first hint
of a more enlightened outlook can be found in a letter from Lady Mary
Wortley Montagu, written in Adrianople in 1717. Smallpox was then
—as indeed it remained until the work of Jenner had produced its
reward—one of the most dreaded scourges of mankind. Every year it
slew its tens of thousands, and those tough enough to recover from an
attack went through life with their faces deeply pitted and scarred.
In this letter we see the beginnings of preventive medicine—the
deliberate choice of a mild form of the disease in order to achieve
immunity against more terrible forms. *How was smallpox first
controlled ?*

LADY MARY WORTLEY MONTAGU

THE PREVENTION OF SMALLPOX

APROPOS of distempers, I am going to tell you a thing
that will make you wish yourself here. The small-pox,
so fatal, and so general amongst us, is here entirely harmless
by the invention of ingrafting, which is the term they give
it. There is a set of old women who make it their business
to perform the operation every autumn, in the month of

September, when the great heat is abated. People send to one
another to know if any of their family has a mind to have the
small-pox : they make parties for this purpose, and when they
are met (commonly fifteen or sixteen together), the old woman
comes with a nut-shell full of the matter of the best sort of
small-pox, and asks what vein you please to have opened.
She immediately rips open that you offer to her with a large
needle (which gives you no more pain than a common scratch),
and puts into the vein as much matter as can lye upon the
head of her needle, and after that binds up the little wound
with a hollow bit of shell ; and in this manner opens four or
five veins.

The Grecians have commonly the superstition of opening
one in the middle of the forehead, one in each arm, and one
on the breast, to mark the sign of the cross ; but this has a
very ill effect, all these wounds leaving little scars, and is not
done by those that are not superstitious, who choose to have
them in the legs, or that part of the arm that is concealed.
The children or young patients play together all the rest of
the day, and are in perfect health to the eighth. Then the
fever begins to seize them, and they keep their beds two days,
very seldom three. They have very rarely above twenty or
thirty [pustules] in their faces, which never mark ; and in
eight days' time they are as well as before their illness. Where
they are wounded, there remain running sores during the distemper, which I don't doubt is a great relief to it. Every year
thousands undergo this operation ; and the French ambassador
says pleasantly, that they take the small-pox here by way of
diversion, as they take the waters in other countries. There is
no example of any one that has died in it ; and you may believe
I am well satisfied of the safety of this experiment, since I
intend to try it on my dear little son.

I am patriot enough to take pains to bring this useful
invention into fashion in England ; and I should not fail to
write to some of our doctors very particularly about it, if I
knew any one of them that I thought had virtue enough to

destroy such a considerable branch of their revenue for the good of mankind. But that distemper is too beneficial to them, not to expose to all their resentment the hardy wight that should undertake to put an end to it. Perhaps, if I live to return, I may, however, have courage to war with them. Upon this occasion admire the heroism in the heart of your friend.

Little progress, however, could be made with man's attacks on diseases until he had a better idea of how they were caused. And it was the work of Pasteur which brought about the great revolution in our ideas on this vital point. *How did Pasteur inaugurate the age of modern medicine?* Some of his early crucial experiments have already been described in his own words, so here we shall call on Lord Lister to answer our questions. No more fitting speaker could be found, for it was Lister who applied Pasteur's discoveries to the art of surgery, and so made operations, which had earlier been a sheer gamble with death, into a sure means of restoring health and prolonging life.

Lister, who was a surgeon at Edinburgh, worried about the large number of operations which "went wrong", resulting in some form or other of blood-poisoning. Anæsthetics were coming into general use and giving the surgeons more time in which to work, and it was most disheartening to find that, the longer and more careful the operation, the greater the danger of blood-poisoning. Lister brooded over this problem, and in 1858 he published a classic paper in the *Philosophical Transactions* showing that inflammation is a reaction of the bodily tissues to some harmful stimulus coming from outside the body. The problem was to find what this harmful stimulus might be, and when, in 1865, his attention was drawn to Pasteur's work on fermentation and putrefaction, he saw at once that this part of his quest was over. All that remained was to find the best means of destroying, or of keeping out, the germs of septicæmia which were floating in the air. So first antiseptic surgery and then aseptic surgery began.

The following speech was delivered at the Sorbonne in 1892 on the occasion of Pasteur's seventieth birthday, when all the most distinguished scientists of the world gathered to do honour to the discoverer of disease germs.

LORD LISTER

PASTEUR AND HIS WORK

I HAVE the great honour of presenting to you the homage of Medicine and Surgery.

Truly there does not exist in the whole world an individual to whom medical science owes more than to you.

Your researches in fermentations threw a powerful light which has illuminated the dark places in surgery and has changed the treatment of wounds from an uncertain, empiric, and too often disastrous business into a scientific and certainly beneficial art. Thanks to you, surgery has undergone a complete revolution which has deprived it of all its terrors, and which has increased its efficacy to an almost unlimited extent.

Medicine owes not less than surgery to your profound and philosophic studies. You have raised the veil which for centuries has covered infectious diseases ; you have discovered and demonstrated their microbic nature. Thanks to your initiative, and in many cases to your own personal work, we already understand the causes of a large number of these pernicious disorders :

" Felix qui potest rerum cognoscere causas ! " [1]

This knowledge has already perfected to a surprising extent the diagnosis of these scourges of the human race, and has pointed out the road which must be followed in their prophylactic and curative treatment.

On this road your splendid discoveries on the attenuation and reinforcement of virus serve, and will always continue to serve, as guiding stars.

As a brilliant example I may quote your studies on rabies. Their originality was so striking in pathology, as in therapy, that many doctors at first regarded you with suspicion. Is it possible, they said, that a man who is neither a doctor nor a biologist can give us information with regard to a disease on

[1] " Happy (is) he who can learn the cause of things."

which the most distinguished lights in medicine have exercised their brains in vain ?

" Quis novus hic nostris successit sedibus hospes ? " [1]

With regard to myself, I recognized too well the brilliance of your genius, the scrupulous carefulness of your instructions, and your absolute honesty, to be able to share such unworthy sentiments for a moment. My confidence has been justified by the event. With the insignificant exception of a few ignoramuses, the whole world now recognizes the grandeur of your achievements in combating this terrible malady. You have furnished a diagnosis which banishes in a moment the agony of uncertainty which formerly tortured one who had been bitten by a healthy dog suspected of rabies. This alone would have sufficed to draw upon you the everlasting gratitude of humanity ; but, by your marvellous system of anti-rabies inoculations, you have known how to pursue and conquer the poison after its entrance into the body.

Monsieur Pasteur, infectious diseases constitute, as you know, the greater number of the maladies to which the human race is subject. You can therefore well understand why it is that, on this impressive occasion, Medicine and Surgery hasten to bring you the profound homage of their admiration and gratitude.

Medical progress since Pasteur's time has been swift and varied, but although it has proceeded along an immense number of most divergent channels, there has always been a certain unity in the methods employed. *What are the foundations of modern medicine ?* Professor A. V. Hill, winner of the Nobel Prize for medicine in 1922, and a leading authority on the mechanism of muscle and nerve, gave a clear and eloquent answer to this question in his address to the British Medical Association at Manchester. Professor Hill shows that nature

[1] " What new stranger is this who has approached our abode ? "

is always making experiments on mankind, and that man's own experiments on nature provide the only sound means of self-defence.

ARCHIBALD VIVIAN HILL

EXPERIMENTS ON FROGS AND MEN

MAN is an inveterate experimenter. Those of us who have been happy enough to see small boys at work, who have been small boys ourselves, or indeed are still small boys, will know what joy is found in taking an old alarm-clock to bits or a bicycle to pieces, in seeing how fast we can run a hundred yards, in breeding rabbits, pigeons, or canaries, in fixing wireless apparatus together, or, when we are older, in trying a new kind of oil or petrol or even a new kind of medicine. Boys and men, however—by which I mean also girls and women—are not the only experimenters to be found in Nature, as any who have watched a kitten or a parrot will know ; and experiments made by monkeys have been scientifically studied. Man, however, is the chief experimental animal, both as experimenter and " experimentee ". Indeed, in many of man's most joyful adventures he acts in both capacities ; he makes experiments upon himself, often to his own great danger or discomfort.

MAN THE EXPERIMENTER

To run in a Marathon race or to try to swim the Channel, to see how far one can ride a bicycle in twenty-four hours, to climb to 20,000 feet, to set out to walk (or to fly) to the South Pole, to make a height record in an aeroplane, to dive under the sea, all these involve trials and experiments upon oneself ; which is one reason why so many apparently useless feats are performed. Every new adventure on which man has embarked throughout the ages, every change in his social and economic and political condition, has meant experiments upon his bodily frame and organization, experiments sometimes successful but often followed by disaster.

In learning the use, treatment, and preservation of food he must, unwittingly often, have made millions of experiments upon himself, thousands of them extremely unpleasant, many of them fatal. Without these experiments, however, the present order of civilization, depending as it does upon a regular supply of food, would have been impossible. When he set out to journey on the sea he experimented on sea-sickness, and later on, as his journeys lengthened, on scurvy and the need of vitamins. When he deserted a natural diet and gathered together in cities he experimented on nutrition and the physiological effects of radiation (or its absence), with rickets as a curious result. When he began to dig deep tunnels, or to work in diving-bells or diving-suits, he discovered that considerations of the physical solubility of gases in his blood and tissues may affect his well-being, and he invented caisson disease. When he climbed high mountains, or went up in balloons, he discovered mountain sickness, and acclimatization to it. When he took to rapid manœuvres in aeroplanes he found out that the human factor is a limiting one, that violent acceleration—" centrifugal force "—may play havoc with his circulation and render him suddenly unconscious. Labouring in hot mines, in extremes of climate, with excess or deficiency of sunlight ; living on sterilized, preserved, or purified food ; breathing quartz dust or carbon monoxide ; working with materials which exert a chronic irritation on the skin, or with ultra-violet light, or with X-rays and radium ; in all such experiments he found limitations to his independence of his external environment ; he made experiments upon himself and others, experiments involving ill-health, disaster, and death to many. Even apart from disease, from the experiments which Nature wantonly insists upon making on us, we cannot avoid making experiments on ourselves if we are to do anything new ; and, even if we do nothing new, we shall probably find we must make experiments still to discover how to remain just as we are.

NATURE THE EXPERIMENTER

I am speaking this evening to those, and the friends and relations of those, who spend their lives in mitigating the results of the experiments which Nature makes upon suffering mankind. Some of these experiments involve bacterial infection and, therefore, are, or will be, to some degree avoidable. Others have no known cause, though some day their character too will be revealed. Others are due to functional disturbances in the mechanism—*i.e.* in the biochemistry and biophysics of the living cells of the body. Some are due to gross lesions which can be seen. Every imaginable ill, great or small, every conceivable torture, physical or mental, you are called upon to witness, and to attempt to alleviate. Some of this suffering is unquestionably due to human folly; some to man's insistence—no doubt for perfectly good reasons—upon living in a civilized industrial State; some to what we call accident —which usually means insufficient skill or a disregard for human safety. Much sickness is avoidable, and *would* be avoided if only questions of public health and infant welfare were taken seriously enough. Many diseases, however, and their consequences, cannot be prevented altogether, until at any rate we know much more than we do to-day. You can predict the course of the disorder when once it is apparent; you can mitigate its evil consequences; sometimes even you can cure it; but at present you cannot tell us how to avoid it altogether.

Now each of your patients represents an experiment performed by Nature, often apparently a cruel and ruthless experiment, and you—her laboratory assistants—are given a variety of them to witness and to try to learn something from. Nature, however, is an extremely bad experimenter; she is, in fact, the imaginary vivisector of anti-vivisectionist literature, whose experiments are made without mercy and without apparent cause. So badly and so casually performed are they, so ill-controlled, that it is often impossible for you to reason accurately from them at all. Small wonder that for thousands of years medical knowledge advanced so relatively slowly, when it had to be based only on experiments such as those which Nature provides. For, as all who have tried to reason from experiments have found, these may be good or they may be bad ; so well made on the one hand, so carefully thought out and prepared before the event, that one may draw sure and decisive conclusions from them ; or, on the other hand, so ill-conditioned, so casually performed, that no certain deduction is possible. In most of Nature's experiments the variables involved are all confused with one another ; the factors at work cannot be disentangled ; half a dozen functions have been interfered with at one and the same time ; the results are not clear cut, and an extraordinary degree of judgment and experience is required before you can reason from them at all.

This cannot indeed be otherwise. To your individual patients you have a duty to perform which is greater than your duty to the rest of mankind. The only way in which the confusion may be avoided is by comparing the results of Nature's casual, random, and complex experiments on human beings with those of simple, properly controlled experiments on living animals. Such a comparison is the means by which in the last few hundred years, and especially in the last seventy-five, medical science has made such startling progress. You can observe, within certain narrow limits prescribed by ethical considerations you can experiment on, man ; the idea of the use of criminals for dangerous experiments, common in the past, is repulsive to-day. In any case, fortunately, the supply of criminals would be inadequate ! To acquire a real knowledge of the factors at work, these observations, these limited experiments possible on your patients or yourselves, *must* be compared with the results of simpler experiments on animals,

in which it is a matter of little moral importance if the patient
dies.

THE EXAMPLE OF CLAUDE BERNARD

Many of you will have read with pleasure and instruction
An Introduction to the Study of Experimental Medicine, by
Claude Bernard. As Lawrence Henderson says in the introduction to the English translation : " The discoverer of
natural knowledge stands apart in the modern world an obscure
and slightly mysterious figure ". To some he is a magician ;
to others he is a sort of stage professor, with blue spectacles
and a long beard, forgetful and untidy ; to a number (including
Mr. Bernard Shaw) he is a criminal or a fool ; to the majority
he is almost entirely negligible, until the newspapers make
some preposterous stunt about him, when he receives unwanted
attention for a few weeks.

> " Whoever fails to understand the great investigator
> can never know what science really is. . . . Not the least
> merit of Bernard's book is that we have here an honest
> and successful analysis of himself at work by one of the
> most intelligent of modern scientists."

I cannot do better than quote what Bernard says about the
experimental method :

> " Man is metaphysical and proud. He has gone so far
> as to think that the idealistic creations of his mind, which
> correspond to his feelings, also represent reality. Hence
> it follows that the experimental method " (by which of
> course Bernard means not merely the *making* of experi-
> ments, which is easy, but the art and science of experi-
> mentation, which is difficult) " is by no means natural
> to man, and that only after lengthy wanderings in
> theology and scholasticism has he recognized at last
> the sterility of such efforts. . . . The human mind has
> at different periods of its evolution passed successively
> through *feeling, reason,* and *experiment.* First, feeling

alone, imposing itself on reason, created the truths
of faith or theology. Reason or philosophy, the mind's
next mistress, brought to birth scholasticism. At last,
experiment, or the study of natural phenomena, taught
man that the truths of the outer world are to be found
ready-formulated neither in feeling nor in reason. These
are indispensable merely as guides : but to attain
external truths we must of necessity go down into the
objective reality. . . . In the search for truth by the
experimental method, feeling always takes the leads :
it begets the *a priori* idea or intuition : reason or reasoning develops the idea and deduces its logical consequences. But if feeling must be clarified by the light of
reason, reason in turn must be guided by experiment."

In these words lies the philosophy of the great experimental
physiologist, whose work and outlook are bearing at last such
rich fruit in experimental medicine. It may seem curious that
the bearing of that fruit has been so long delayed. Henderson
is probably right in seeing in the growth of bacteriology,
following on the discoveries of Pasteur, the cause which drew
men's attention away for a time from the more fundamental
study. The magnificent edifice of bacteriology is not yet
complete, but one's mind's eye can grasp already what its
dimensions, its plan, its proportions, its significance, are to be.
Thought is returning to Bernard's conception of medicine as an
experimental biological science, of which, as he rightly said,
zoological vivisection is, and is likely to remain, an integral part.

Let me take an example from Bernard's own work to illustrate how a few controlled experiments on living animals may
shed light on the countless uncontrolled experiments of Nature
on living man. The existence of vasomotor nerves,[1] the
control of blood-vessels by the innervation of the muscle-
fibres that lie around them, are matters of ultimate importance,
not only in our understanding of bodily functions, but in

[1] *I.e.* nerves controlling the supply of blood to the brain, stomach, liver, skin. or
any other part of the body.

dealing with the phenomena of disease. In 1851 Bernard made his first communication on the effect of dividing the cervical sympathetic nerve in the neck of a living animal. The ensuing rise of temperature on the affected side, a surprising and quite unexpected phenomenon, led by a long series of researches directly to our knowledge of the vasomotor system. It is almost impossible, as Michael Foster says, to exaggerate the importance of the result, the influence which it " has exerted, is exerting, and in widening measure will continue to exert, on all our physiological and pathological conceptions, on medical practice, and on the conduct of human life. Whatever part of physiology we touch, be it the work done by muscle, be it the various kinds of secretion, be it the insurance of the brain's well-being in the midst of the hydrostatic vicissitudes to which the changes of daily life subject it, be it that maintenance of bodily temperature which is a condition of the body's activity : in all these, as in many other things, we find vasomotor factors intervening." In inflammation and in fever, in shock, in any of the disordered physiological functions which constitute disease, whatever be the tissue, vasomotor influences have to be taken into account.

All this dominant knowledge has come from Bernard's initial experiment in cutting the cervical sympathetic nerve. A simple experiment on a living animal suddenly brought a great light into a field where men had been groping in vain with the help only of clinical observation. The result of the experiment was the first clear light which broke upon the subject : and it was the following up of the teaching of the experiment which supplied the interpretation of the hitherto obscure clinical facts.

THE VALUE OF ANIMAL EXPERIMENTATION : HARVEY AND BERNARD

We hear so much objection raised to-day to the use of living animals, for experiments designed to solve the problems which medicine and physiology supply, that one must continue to insist that nearly all fundamental knowledge in the medical sciences has, in fact, arisen from such experiments. In the dedicatory epistle of *De Motu Cordis*, addressed to that " most excellent and most ornate man ", the president of the College of Physicians in London, William Harvey, speaking of the experiments on living animals by which he had for many years demonstrated the fact of the circulation of the blood, remarked :

> " Neither do Philosophers suffer themselves to be addicted to the slavery of any man's precepts, but that they give credit to their own eyes ; nor do they swear allegiance to Mistress Antiquity, as openly to leave their friend Truth. For as they think them credulous and idle people, who at first sight do receive and believe all things, so do they take them for stupid and senseless that will not see things manifest to the sense, nor acknowledge the light at midday.... Likewise all studious, good, and honest men do never suffer their their minds so to be o'erwhelmed with the passions of indignation and envy, but that they will patiently hear what shall be spoken in behalf of the truth or understand anything which is truly demonstrated to them ; nor do they think it base to change their opinion, if truth and open demonstration so persuade them."

If only the precept of these words could be heeded by those who object to the experiments of Harvey and his successors !

When Harvey, as he said, " first applied his mind to observation from the many dissections of living creatures as they came to hand " to find out the nature of the motion of the heart, he " straightways found it a hard thing to be attained, so that he could almost believe that the motion of the heart is known to God alone ". At last, however, " using daily more search and diligence, by often looking into many and several sorts of creatures, I did believe I had hit the nail on the head and thought I had gained both the motion and use of the heart, which I did so much desire ".

And so, by the vivisection as he says of " toads, serpents, frogs, house-snails, shrimps, crevisses, and all manner of little fishes ", together with eels, dogs, swine, doves, chicken embryos, crabs, wasps, hornets, gnats, bees, geese, rats, sheep, adders, lice, swallows, partridges, hens, and swans, modern physiology and modern experimental medicine were born. For those who, in Harvey's words, are not " stupid and senseless ", who are ready to " acknowledge the light at midday ", we have here one of the greatest discoveries of all time ; a discovery greater even than that of the circulation of the blood ; the discovery that by comparative experiments on " many and several sorts " of living creatures we may reveal the nature and working of man's body, we may increase his power and happiness and wisdom, and cure and protect him from suffering and disease.

This lecture has been entitled " Experiments on Frogs and Men ". The physiologist's little friend the frog has been chosen as the general type of the experimental animal, and I have just read you a list of all its colleagues which William Harvey employed in his discovery of the circulation of the blood. In using these various animals he assumed, unconsciously, what we now know to be a fact, the fact (I do not call it the theory) of evolution. He realized, being a sensible person, that the hearts of snakes or crabs, of snails or fishes, do not differ so much from those of the higher animals and man that an investigation of the former would throw no light upon the latter. There *are* differences, of course ; no reasonable man supposes that by studying the circulation, or, indeed, any function, in the frog or even in the dog alone, we shall learn enough to make us competent physicians. The differences, however, are often more apparent than real ; the fundamental properties of the ultimate living units—that is, of the single cells which build up the various organs— are surprisingly alike. By the methods of comparative physiology, or of experimental biology, by the choice of a suitable organ, tissue, or process, in some animal far removed in evolution, we may often throw considerable light upon some function or process in the higher animals, or in man.

THE FURTHER RANGE OF DISCOVERY : DEFICIENCY DISEASES

I have spoken of two discoveries of the more distant past, those of Harvey and Bernard, which were made by the aid of experiments on animals, discoveries on which the practice of medicine to-day is based. Where would you be if you did not know or would not acknowledge the circulation of the blood ? How would your patients fare had you never heard of the control of blood-vessels by nerves ? To refer to other discoveries, how safe would surgery be to-day had the researches of Pasteur on animals and their application by Lister on men never been made ? When you take a blood pressure, do you or your patients give full credit to the Rev. Stephen Hales, who laid the basis of that particular science, by experiments on living animals performed in the parsonage at Teddington ? Do those whom, by a miracle of skilful surgery, you save from the misery of hyperthyroidism and restore to normal health realize that our knowledge of the thyroid gland and the possibility of this particular operation are due alike to the so-called vivisection of animals and men ? Brain surgery, renal surgery, the surgery of the nerves, the spleen, the pancreas, the stomach, the ligature and suture of arteries, the use of artificial respiration, the treatment of anthrax, rabies, diphtheria, tetanus, and syphilis (to name only a few) ; the researches which led to all these were intimately bound up, from first to last, with experiments on, with the vivisection of, animals as well as men. This is a matter—as Foster said about Bernard—of plain historical fact, which those who wish may verify. No manner of abuse or misinterpretation will get over a plain historical fact. Is there a fool so great, or a criminal so wicked, that he would be willing now to *cause* the

suffering to men and women which these advances in medical
and surgical knowledge have already saved—and will save
in the next 50 million years—in order to *prevent* the suffering
to animals, such as it was, by which, once and for all, they were
achieved ?

In 1889 Mering and Minkowski found that complete
removal of the pancreas in animals is followed by severe and
fatal diabetes, and a long series of experiments led to the
conviction that some chemical substance, prepared by certain
patches of cells, the so-called islets of Langerhans, existing
in that organ is essential to the normal utilization of sugar
by the body. So certain did this seem that—although all
attempts to isolate it had failed—the name " insulin " was
given to this hypothetical substance by de Meyer in 1909,
and independently by Schafer in 1916. I need not tell you
how the Toronto workers in 1922 succeeded in isolating this
substance, and how effective it has proved in the treatment of
human diabetes. Those of us with friends and relations who
are maintained in practically normal health by the daily
administration of insulin, and who—in all human certainty—
would die in a few months were the supply of insulin prevented,
can appreciate what experiments on animals have done. Not
only did the discovery of insulin rest absolutely upon experiments on dogs, rabbits, and other animals, but even its supply
cannot be maintained at present, until scientific knowledge of
its nature is advanced much further, without continual
experiments on mice and rabbits, for its standardization.
This again is plain historical fact. There are people who
assert, and affect to believe, that insulin has no effect on
diabetes, or that, in fact, it does harm ; they base their assertion on the crude handling of statistics. There are still people
who believe that the earth is flat, and that spirits can make
images on photographic plates. Have those who protest
against insulin ever seen a patient, sick unto death in diabetic
coma, pulled as by a miracle out of his sickness and restored
to health ? Would they dare to condemn to death a healthy

happy man, in active work, a man such as many of us know, but a complete diabetic, by refusing him his daily dose of insulin because it has to be tested and standardized on rabbits or mice ? The question really needs no answer ; ultimately these are humane people, though they have bees in their bonnets.

In 1926 the experiments of Whipple on the regeneration of blood-corpuscles by dogs, bled and then fed on liver, led to the discovery that pernicious anæmia can be averted, if not cured, by feeding human patients with the same substance, or by treating them with liver extract. It is healthy, I know, to have a certain mistrust of " experts ", but it is healthier still to have a mistrust of fools. Those who are qualified to judge, and have seen the evidence, are convinced of the efficacy of this treatment ; they have no shadow of doubt about it, and it is plain historical fact again that it was led to by experiments on animals. Unfortunate, maybe, but true ! Yet a medical man I once met, a valued member of an anti-vivisectionist society, asserted in my hearing that the liver treatment of pernicious anæmia is a delusion, since in fact there is no such disease at all ; according to him it is really cancer of the bone-marrow—whatever that means (and he himself has a cure for cancer). Whose opinion should the public accept in this matter, which after all is not altogether unimportant, since it involves quite a number of human lives ?

We hear a great deal in these days about vitamins : the public enjoys the word, especially in America. I once had the privilege of sitting at dinner between two distinguished members of the advertising profession. One, I believe, was an arch bill-poster, the other I found later was concerned with the advertisement of patent foods. Neither knew who I was. The first began : " When I sit down next to a stranger I generally open the conversation by asking him, ' Do you know my distinguished cousin ? ' "—naming one of the greatest of living investigators. I did, and conversation progressed favourably, finally leading to vitamins. When I turned to my second neighbour I told him how conversation had opened on my other side, and how, of course, it had turned to vitamins. He replied that vitamins were the curse of his existence, that he wished they had never been discovered, that their discoverer was a criminal and ought to be locked up, and so on. He seemed in other respects a reasonable and healthy man, so I made further inquiries. It appeared that in every patent food vitamins have to be present—at any rate in the advertisement—and it gives him sleepless nights wondering how the word (not necessarily the substance) can be introduced. . . .

The discovery of vitamins was made by experiments on animals—though I admit that scurvy was discovered inadvertently, and a cure for it invented, by experiments on sailors. My friend, Professor Elliott, once said to me that to Hopkins, for his work on nutrition and accessory food factors, medicine would owe more than to anyone since Pasteur. I at any rate can testify that Gowland Hopkins' researches involved experiments on animals—did I not work for several years in a coal-cellar at Cambridge, in the company of several hundred of his rats ? I never saw him maltreat them—in fact, I have good reason to believe him to be the most kind-hearted of men—but this is " vivisection ", to give it its legal title, and Hopkins' discoveries were made by " vivisection " in a coal-cellar. I am happy to have lived among the rats of such a " vivisector ", and to have seen such a discovery emerge.

These matters are not unimportant. Human lives and human happiness are involved. Look up the history, recent or remote, of medical knowledge and treatment, explore the story of any medical discovery, and you will find experiments on men and experiments on animals inseparably mingled. It cannot, and it should not, be otherwise. *Anti-vivisection, in fact, is anti-scientific medicine* ; and those who work in laboratories need a fearless recognition of the fact by those who work on patients. Your rewards, social and financial, are not usually less than theirs ; it is you who have the ears of our rulers, whether in cottage or in mansion ; they remain, to the public, obscure and somewhat mysterious people ; it is you, in fact, and not they, who can stop, if anyone can, the crime of hindering or preventing medical progress by means of scientific experiment. The object of our opponents is frankly to put an end to medical research, in so far as animals are used, instead of men, for experiment. They say so openly. You reply that this is inconceivable. It is not inconceivable ; it might very well happen. There are States in America where even the teaching of evolution is forbidden. Do not let England become the laughing-stock of the world, like Tennessee !

Experiment the only Way

Is it indeed conceivable that we can find out about life without examining life ? The whole history of human knowledge and progress tells us the danger and folly of scholasticism and unverified hypothesis, the advantage of unbiased experiment and observation. If you wish to find out about the stars you must observe the stars, you must experiment with the materials which you suspect the stars to contain, you must try to reproduce, and to experiment under, the conditions which obtain in the stars ; you must study spectra, electrons, radiation, gravitation ; you will not be content merely with pictures of the sky at night. If you want to solve the problems of flight you must experiment on flight ; you will never solve them merely by examining dead birds or fossil birds, or even living birds in cages. If you want to understand the properties of matter you must make experiments with matter, and apply mathematical calculations to your results as necessary ; it is not sufficient to read what the Greek philosophers said about matter. If you want to know how living things evolved you must examine the strata of the earth, in which you can find them in their various stages of evolution ready to look at. You must not be content with the first chapter of Genesis. If you want even to play tennis you must experiment in playing tennis—all the books in the world will teach you nothing without the practical experience. So, if you want to understand life and the problems involved in life, if you want to acquire the power which an understanding of living processes will bring you, you must observe living things, you must take them to pieces to see how they work, you must become familiar by long experience with the way they behave.

Living processes, as Bernard upheld, are subject to the same determinism as the phenomena of the inorganic world. Cause and effect are related in the same way. There are no spooks, no spirits, no magic, no supernatural agency, no lack of causal relationship to hinder us from applying the same kind of exact quantitative experimental analysis as men are accustomed to apply to the facts of physics, chemistry, engineering, or astronomy. The only difference is that the problems of physiology, taken in its widest sense, are far more difficult and complex than those of the so-called exact sciences. The problems of biology require greater and not less experimental skill, more patience and not less, more time to unravel, more judgment and more understanding for their solution. Biology is *not* a soft job, for people with second-best brains ; though physicists and chemists sometimes affect to think so. There is no evidence, however, that its problems are not ultimately soluble. We may not solve, as the newspapers would say, the " mystery of the origin of Life " ; any more than astonomers can tell us how the universe of matter, energy, and radiation came into being in the beginning. Probably there *was* no origin and there *was* no beginning ; at least there is no evidence at all on the subject. The real difficulties in biology are great, but they are experimental difficulties, not philosophical ones. If we wish to understand life we must experiment with it, not talk about it. Our experimentation, of course, must be led to by hypothesis, for random experimenting is as useless, as dangerous as unverifiable hypothesis ; but when once we have appealed to Nature to judge between our theories, we must accept her decision, without bias of hypothesis, or philosophy, or religion. We must trust absolutely to experience, guided by reason. *We must admit no challenge to the arbitration of experiment, except better and better and still better experiment.* The field of biology is vast, much of it is still quite unexplored, but we have a few landmarks to guide us. The fact of evolution is certain, the fact of determinism, the fact that the laws of the inorganic world apply in biology with the same precision as elsewhere. The field is not ripe yet for much more theory— exact quantitative experiment on living creatures and tissues is the present need, and will remain so for many years to come.

How, then, shall we proceed ? By experiments on man ? Surely, for man is the only experimental animal who will co-operate fully with the experimenter, and most important results have been obtained in recent years by controlled experiments on man. Besides, the habit of experimenting on man, on themselves and their friends, is a valuable lesson to learn for all those whose lives are to be devoted later to medicine ; and no less important should it prove in connexion with the scientific study of industrial and social welfare. Let us, however, insist and continue to insist that human physio- logy and medicine are branches of biology as a whole. Man has a kinship, through evolution, with all the plants of the earth, the beasts of the field, the birds of the air, the creatures of the sea. Fifty years ago this conclusion was opposed by all the resources of sentiment and of organized religion. Even to-day there is widespread objection among the uneducated to the teaching of evolution, but none on the part of those who have studied the evidence. We *are* akin to the other creatures, and a frank acknowledgment of the fact need not, in fact does not, diminish our appreciation of the nobility of human nature or of the sanctity of human life. One objection to the use of animals in experimental medicine is undoubtedly the same old anti-evolutionary prejudice. To- day, however, the conclusion cannot seriously be contested ;

Bernard is admitted to be right, medicine and human physiology *are* branches of experimental biology. We must make them so in fact as well as in theory, in hospital and laboratory as well as in lecture-room.

Let us advance, therefore, by experiment, hopefully and without fear. Harvey, by his experiments " using daily more search and diligence, by often looking into many and several sorts of creatures ", did finally " hit the nail on the head " and so laid the basis of modern scientific medicine. To-day there are unlimited opportunities for experimental biologists, adequately schooled and disciplined in hard reasoning and exact methods, in its border sciences. From frogs to elephants upwards, from frogs to filtrable viruses downwards, we may take our choice, according to the demands of the experiment ; and the end of it is man. Puny, fearful, unreasonable, selfish he may sometimes be ; yet possessing always elements of nobility, courage, kindness, and devotion such as no other animals possess, which make him, as was said 1900 years ago, " of more value than many sparrows ". On his behalf, for his health, his happiness and his wisdom, for his freedom from suffering, superstition, and ignorance, experimental biology and experimental medicine should be ready to sacrifice, if needs be, many sparrows and many frogs.

Now let us look at what is undoubtedly the commonest epidemic illness of to-day. Most people have had an occasional attack of influenza, and since the sudden and violent outbreak of 1918 this disease has made periodical returns—fortunately in a much less virulent form. *Can influenza be reduced ?* An interesting talk on this subject was given early in 1951 by Dr. C. H. Andrewes, who is in charge of the World Influenza Centre at Mill Hill.

CHRISTOPHER HOWARD ANDREWES

CAN INFLUENZA BE REDUCED?

WE in Britain are at present in the middle of an influenza epidemic. Thousands of people have been
in bed and away from work ; some, mostly those who already
had some other illnesses, have died. What causes this influenza ? Why does it come every so often ? Is anything
being done about it ?

There is quite a lot to tell. First, influenza is a disease on
its own, not a severe kind of cold, and it is due to a virus.
Viruses are the smallest sort of germs : it is viruses which
cause measles, smallpox and poliomyelitis. Viruses grow only
in living cells. They have therefore to be studied in experimental animals (ferrets are used in work on influenza) or,
as is being done more and more, by growing them in fertile
developing hens' eggs. The eggs are incubated till the embryos are a fair size and then a window is made in the shell
at the broad end of the egg. Virus is obtained for study by
getting a patient to gargle with salt solution and spit the
garglings out—these we call " throat-washings ". These
throat-washings from a case of influenza are introduced into
the little bag of fluid surrounding the embryo. After a few
days' further incubation, the fluids are removed and some
fairly simple examinations on the laboratory bench allow one
to say if influenza virus is present, and if so to which of the
main types it belongs. There are two main types of virus.
We call them A and B. They both cause the same kind of
illness, but an attack by one does not give you even temporary
immunity against the other.

Influenza only visits this country roughly every other year.
People who say they have influenza every year or oftener, do
not get real influenza every time ; more probably they respond
in an unusual way to the common cold or some other infection.
But you may have two attacks of influenza in one year, one
due to virus A, and one due to virus B, but not more. The

influenza virus grows in people's throats and it is carried in the air on minute droplets coughed and sneezed out. If you breathe in some of these droplets you may go down with 'flu. In the course of your attack you develop antibodies against the influenza virus in your blood : that is to say, your blood acquires the power of knocking the virus out.

During an epidemic more people escape infection than catch it. I had better say " seem to escape infection ". Examination of blood samples taken at the tail-end of an epidemic from people who have *not* had 'flu shows that a lot of them have developed these antibodies, though they had no symptoms of 'flu at all. We have evidence which leads us to believe that after breathing in a small dose of 'flu virus you may acquire a resistance quite painlessly and comfortably (as it were) without any symptoms ; and then later on you can inhale a bigger dose of germs with impunity. If we could only ensure that everyone would receive only such small doses, 'flu would be no longer a problem. At present the best hope of passing unhurt through an epidemic is to avoid crowds as far as you can and trust that you will then breathe your 'flu germs in by ones and twos instead of by hundreds and thousands.

That is (shall we say ?) the consumer end of the story of infection. The producer or distributor end is this : the man with 'flu who does not go to bed, who is a martyr and carries on till he drops, is doing his best to ensure that his fellow-man takes in hundreds of germs and so inevitably catches 'flu. One may add here that good ventilation helps to keep the 'flu germs in the air down to a low level.

Is there anything else that can be done besides avoiding infection ? It is possible to make vaccines against 'flu, and they are made from the fluids of the hens' eggs in which the virus is grown. The vaccine can be injected under the skin and definitely gives some protection against the corresponding virus. But there is a difficulty about this. The 'flu virus keeps changing ; the fellow causing the last epidemic is not necessarily the same as the present villain, and the vaccine made to thwart the last one may be useless in the present outbreak. These new variants of the virus which keep turning up can be traced as they spread from country to country. If we could only spot, ahead of time, what sort of virus was coming to us, we might be better able to cope with it when it arrived. This problem is the special job of the World Influenza Centre; this is an activity supported by the World Health Organization and it is now situated at the National Institute for Medical Research at Mill Hill. Here we keep contact with laboratories all over the world studying influenza; and they send us information and strains of virus. These either come by aeroplane packed in ice; or else they are completely dried and sent by post in sealed glass ampoules. In one morning we received some from Turkey, from Yugoslavia, and from France.

In 1948-49 we were able to follow the spread of an epidemic which began in the Mediterranean area in Sardinia. It spread from there to continental Italy, up to Switzerland, Austria, France, and northern Spain, and to western Germany and Holland. About the New Year of 1949 it reached England but was not so extensive here as on the Continent; it finally got as far as Denmark and Iceland. We received virus from most of the countries in the track of the epidemic and they were all alike, a variant of the most important influenza virus, virus A. These studies showed that the spread of new strains of virus from country to country is a real fact and not just an old wives' tale.

But from the late spring of 1949 to a few months ago, it would have been a matter of the greatest difficulty to find one single virus of influenza in this country, or indeed in western Europe, outside a laboratory. It is like the puzzle of where the flies go in the winter. We are pretty sure the virus lurks somewhere between epidemics but we just have not a clue as to where. We did have, it is true, a certain amount of influenza a year ago, but that was due to the other kind of influenza virus, virus B, which is never so troublesome as A. Virus A itself put in just a brief appearance last June in Sweden around Stockholm, causing quite limited, local outbreaks. All the same, this little flurry of influenza interested us quite a bit. On several previous occasions 'flu has caused a little local trouble in the early summer, and has then disappeared only to turn up in the same general area and start a widespread outbreak in the following autumn. So we kept a special eye on this small outbreak in Scandinavia. We even made a small batch of vaccine against the Swedish virus. Nothing happened till November. Then 'flu broke out in several districts in Denmark—the same strain of virus as in the summer—and soon after in northern Sweden and Norway. Since then it has appeared in other parts of Europe. The fact that its first appearance in England was around Newcastle suggests that it may well have reached us from Scandinavia. So far as studied, this season's viruses are not very different from the 1949 strains. During last December 'flu broke out in two other countries—in Northern Ireland and in northern Spain ; these may well have been separate points of origin, unconnected with the Scandinavian virus. One can visualize epidemic ripples spreading out over Europe from these three separate centres—Scandinavia, Ireland, Spain. We shall be able, we hope, to discover more about what has been happening when this epidemic is over and we have been able to compare all the various strains and perhaps plot their movements.

One object of all this is to find out how to forecast what strain is coming to a particular country and when, so that one can be ready with a vaccine. Many people, especially journalists, come out, when they meet us, with two stock questions : first, " Have you isolated the virus ? ". " If so, have you prepared a vaccine ? " A vaccine may be very useful, but it is not the only or necessarily the best answer to our problem. In a big epidemic, more than half the country may be exposed to 'flu : think of the expense and man-power involved in protecting everybody by making some millions of doses of vaccines out of eggs ; and, after all that, an unexpected brand of 'flu may turn up, so that millions of doses of vaccines might be wasted. You may see now why our plan of attack on 'flu is not only to find better vaccines but is also a long-term programme designed to understand more about the behaviour of the virus itself—why and how it varies and changes, where and how it disappears from sight for months and months, and above all, how it reappears and gets going again. I suspect that the conditions under which it becomes able to raise its head again may be rather special ones. If we knew what they were, we might be able to scotch a budding outbreak at the source much more cheaply and efficiently than by vaccinating every Tom, Dick, and Harry in the country.

The particular influenza now afflicting us is more than a nuisance ; it is killing some people and causing great loss to industry. But it is relatively mild. At the back of the minds of many of us is the memory of the great world-wide influenza epidemic of 1918 and 1919. In the excitement of finishing up the first world war that epidemic did not attract as much attention as it might have done; but it was in fact the greatest pestilence which has ever afflicted mankind. It killed in a few months more than were actually killed in all the fighting of that war. Is it possible that the rather mild 'flu of these times could turn into the fatal 1918 type of disease ? We cannot deny the possibility ; we can only say that we see no signs of it.

The feature of that epidemic was that the disease was fatal to young previously healthy adults, while old, infirm people got through pretty well. The reverse is the case at the present time. A week or two ago rather disturbingly large numbers of deaths were reported from Merseyside. But many deaths always occur during 'flu outbreaks and they are chiefly in people who already have chronic illness, and who are not likely to live very long in any case. That is the situation in the present epidemic, a very different picture from that of

1918. At that time the virus went, as it were, into partnership with a number of larger germs, with the streptococci of sore throats, the staphylococci of boils, the pneumococci of pneumonia, and other bacteria. It was mainly these virus-bacterium partnerships which killed people in 1918. But that was before sulpha-drugs, before penicillin, streptomycin, and the other weapons of defence we now have in our medicine chests against the bacteria. So perhaps we shall never again have such a killing plague as in 1918. We prefer, however, to leave nothing to chance if we can help it. We shall go on studying the variations of the 'flu virus, the way it appears, the way it spreads and disappears ; we shall try to find out where it is hiding when we have lost track of it ; we shall go on trying to improve vaccines and we shall look for what we very badly need, a drug active against viruses. Penicillin and other new drugs are marvellous against most bacteria but not against viruses. If only we had a drug to scotch *them* we could feel that man was in sight of winning the last big battle in the war against the germs of disease.

Our manners have no doubt improved considerably in the course of the past century, but coughs and sneezes without benefit of hand-kerchief are still painfully common in public halls and vehicles. Until this practice comes to be regarded as just as bad as spitting we shall have to put up with the usual increase in airborne diseases every winter. Public health, indeed, depends to a very large extent on the public, and a little thought and consideration for others could make a big improvement, at least so far as coughs and colds and 'flu are concerned. *How does public opinion affect the nation's health ?*

The answer is supplied by Sir George Newman, who was Chief Medical Officer of the Ministry of Health from 1919 until 1935. In it he outlines the chief factors in maintaining good health, and also the four great defences of the body against disease.

SIR GEORGE NEWMAN

PUBLIC OPINION IN PREVENTIVE MEDICINE

SOME kind of public opinion has no doubt existed in the world from the earliest history of mankind. Yet it is only in recent times that such opinion has become influential on a large scale. Sir Robert Peel described it in 1820 as consisting of a " great compound of folly, weakness, prejudice, wrong feeling, right feeling, obstinacy, and newspaper paragraphs ". Whilst we may not be disposed to assent to all the terms of this definition, it contains certain ideas which are characteristic of public opinion in our own day.

For when we come to analyse it we shall find that, like the history of the English race, it is a mosaic, an amalgam, of sentiment and of reasoning on available data. Lord Bryce has furnished an instructive discussion on the nature and action of public opinion, in which he suggests that sentiment is contributed most largely by the mass of the people, and that reasoning on available data combined with self-interest is characteristic of the more educated classes. No doubt also we must add to these two elements a certain amount of prejudice, of respect for authority, and of facilities for expression. Indeed it is probable that, from the time of the Reform Bill of 1832, public opinion in England has been awakened, kindled, and become operative by means of public agitation, the enormous development of public speaking, the freedom of the press, and the ever widening education of the individual. It is not the local Council, it is not Parliament, it is not even Government that is the ultimate authority—for that rests with the aggregate of individual citizens. They are master—Government is their servant. Whatever be the factors, and they are numerous and complex, the world is moved to-day, as never before, by the indefinable power of public opinion, and Governments, as well as habit and custom, are impelled or moulded by " the man in the street ". All through the broad territories of the British Empire public opinion is paramount, and government is by the people.

The practice of preventive medicine in its modern meaning rests upon the growth of medical science and the application of that knowledge to the problems of disease. During the last half-century the increase of physiological and pathological knowledge, including that of infection, has been one of the outstanding features of the age in which we live. We now know two certain facts about disease : first, that it is not something arbitrary, capricious, occult, or accidental, but an effect of definite causes and conditions ; and secondly, that these causes and conditions are in large and increasing measure controllable by man. For the greater part of the nineteenth century we did not know the cause of leprosy, typhoid, tuberculosis, diphtheria, cholera, or plague, and the process of their control had no basis in etiological fact. To-day we know the fundamental truths of causation, and therefore, for the first time, public and personal health has become purchasable—but the purchase involves desire to purchase, understanding of what is to be bought, and adequate resources of knowledge and of money. There are two things we desire to purchase, a healthy life and a long one. In other words, we seek to reduce, and if possible abolish, invalidism and physical disability, and to postpone the event of death. This, in a word, is the business of preventive medicine. It is to make human life better, larger, more capable and useful, happier—and it is to prolong our days. Its purpose is to make the time in which we live, and the future, a better time for all men. . . .

The way of health is not hidden in abstruse science. For wisdom and understanding in this subject grows and develops, partly, it is true, out of the growth of medicine, but partly also—and in no mean degree—out of experiences, habits, aspirations, and morals of the people as a whole, evolving slowly out of their communal consciousness rather than being superimposed, ready made, from above. For medicine is human in application as well as technical ; it is social in purpose as well as scientific in method ; it is moral in claim as well as intellectual. Only a people clean in mind and body, within and without, can withstand " the pestilence which walketh in darkness " ; and thus the social and moral standard of a people, its national character, bears relation to its health, and that, and not the medical issue alone, is the decisive factor.

The elements of health for the body are nutrition, fresh air, and exercise. Cleanliness, warmth, and rest are also necessary, but they are secondary and consequent upon the other three factors. The fundamental requirements out of which emerge all other necessities are these three. If these are present and adequate we have something approximating to full life ; if they are absent or inadequate we have insufficiency, poor physique, disease, and even death. Though this knowledge is as old as the history of the human family, it is still partially applied to the building of men or the rearing of a race. For though the general proposition is simple, its application is complex. What is the ideal form of nutrition in any given climate or for any given age of life ? How can we live in the open air if we also live in houses and in cities ? By what means can we secure sufficient exercise, and of what nature should it be, and how can we escape the condemnation by Galen of the over-specialization of Greek physical culture ? The answer to these questions comes only by knowledge. For, given a " living wage " and given a sufficient yield of the proper food necessary to man's health—and, speaking generally, these desiderata are available in this country—there can be but one answer : people do not live the healthy life because they lack knowledge. It is ultimately a matter of experience and of knowledge. As a nation we shall never win through to a high physical standard until the great mass of the people are educated sufficiently to be able to choose the way of health. It is a practice and not a theory.

Let us consider this issue for a moment. It is common

knowledge that man's diet should be mixed, varied, sound in quality and sufficient in quantity, consumed at fairly regular intervals, and appetizing and digestible. It should contain an adequate supply of protein, fat, carbohydrate, inorganic salts, and water—and these substances are represented in greater or less degree in the common articles of diet, meat, milk, bread, sugar, starch, rice, oatmeal, etc. Further, it is now established that, in addition to these necessary constituents, certain unidentified living principles, known as accessory food factors or vitamins, must be present in the diet to ensure health and nutrition. These bodies are contained in peas, beans, eggs, animal fats, milk, butter, cod-liver oil, green vegetables, potatoes, fruits, lemon juice, and so on. If a chemically ideal and model food be prepared in the laboratory with the proper amounts of each essential constituent, and then it be sterilized, dried, or otherwise " preserved ", it may be deprived of some or all of its vital principles. Hence, dried foodstuffs, preserved vegetables and tinned meats, though possessing some practical advantages, are reduced in value as complete foods. They become auxiliary only, for they are deprived in some measure of their vital elements. Moreover, nutrition does not consist only of pabulum, the food. There must be healthy activity of those physiological processes which have to do with mastication and preparation, with absorption and assimilation, metabolism and excretion. Healthy and complete nutrition is infinitely more comprehensive than mere feeding, mere filling of the stomach. It connotes a healthy body in all respects, a brain and nervous system in tone, a healthy muscular and digestive system, circulation of blood and lymph. Now, when we turn to the dietetic conditions of the great mass of the workers we find a tale of beer and bread, of tea and pickles, of tinned meat and cakes, of a bit of bacon and a piece of cheese—and of an unstable digestive system and an impaired physique. Variety, appetizing cookery, freshly prepared food, the healthy conditions of sound digestion—these essential things are often absent.

" Experience in Manchester teaches that the most fertile cause of ' poor physique ' in children is the gross ignorance of the simplest form of domestic economy and cookery amongst the mothers. . . . The difficulty was not due to poverty (except in a practically negligible number of cases), but to the lack of knowledge of the mothers (who could have afforded to buy the necessary food) as to how to cook it. It would appear that the scientific method to deal with the question of ' poor physique ' is to begin by due instruction of the elder girls in the schools and of the young women in the continuation classes in artisan domestic economy."

Medical officers have been saying this for a generation, and happily the provision of school meals by local education authorities, the canteen movement in factories, and the national movement for a reformed milk supply, are at last beginning to have effect. But the real issue is still the education of the people. By our ignorance and our willingness to be ignorant, both in theory and in practice, we provide the conditions which inevitably lead to the growth of an enfeebled race.

All this applies equally to the necessary conditions of health in respect of fresh air and exercise.

" The national adoption of the open-air school," wrote Dr. Leonard Hill " the garden city, the open-air factory, and open-air exercises, cannot fail immeasurably to improve the health and the enjoyment of life of all concerned. There could be no more far-reaching or beneficial reconstruction work done at the present time, both for the health of the child and the adult, than the introduction of what I may call the practice of the open-air life both in school and in factory. . . . It is sunlight, coolness, dryness, diversity of cooling effect and movement of the air, which will exert a health-giving effect."

The physiological advantages of adequate and continuous oxygenation of the blood are incalculable—it is the essential condition of conveying life-giving oxygen to all parts of the body—but these advantages are not secured by ill-ventilated schools and factories or by stuffy dwelling-rooms. With few exceptions, the means of ventilation are there ; what is lacking is the maintenance of the means by the individual ; and neglect of maintenance is due to lack of knowledge and to the dominance of bad habit.

In order to build up a sound physique the nation also needs to have available a complete scheme of educational and recreative gymnastics, that is, a system of carefully chosen, graduated exercises, designed on physiological principles, to train both body and brain, and combined with games, swimming, field sports, and dancing. . . .

It is not enough for 40,000 youths to watch a football match or a boxing competition or a horse race. They also must play the game. It is not enough to make provision for organized games and physical development in the schools for children up to the age of 14 only. It is necessary to carry this on right through adolescence and into adult life. What was done for the soldiers in France can equally well be made available for the industrial workers at home. The evidence before the Ministry of National Service was cumulative and convincing that hundreds of thousands of the men rejected for military training on grounds of ill-health had been debarred by social circumstances from any recreational exercise whatever.

> " The health-giving recreations of the industrial classes are almost nil " said the Ministry of National Service " their spare time being taken up in parading the city streets, attending picture-houses, and watching certain games instead of taking part in them."

There are few problems more pressing in this country at

the present moment than the provision of facilities for games, athletics, baths, swimming, and all forms of healthy physical recreation. If we do not solve this problem, at least for the great industrial towns, we must not expect anything but low physique in the population. Nor can we afford to forget the women—the source of the new race. If music and dancing, golf, hockey and tennis, and recreation in the open air be good for any young woman, they are good for all. We do not want to train our people as acrobats for the prize ring or the Olympic contest, but it is imperative that we should equip them for an all-round healthy life. When as a nation we secure for the people as a whole the simple elements of nutrition, of the open-air life, and of adequate exercise—so profound will be the effect and so great the example—a new day will have dawned for mankind.

The second reason for education of the public is the prevention of disease. Clearly, one of the essential methods of prevention is a high standard of personal health, for sound physique is more resistant to infection and the onset of disease than poor physique. Everything therefore which tends to personal health is a factor in prevention. But more than this is required.

As we have seen, invalidism, disease, and premature death are due to a relatively small number of morbid conditions. Respiratory disease, indigestion, and general minor ailments (debility, neuralgia, lumbago, septic conditions, minor injuries and accidents, skin disease, anæmia, etc.) constitute 70 per cent. of insurance cases, more than a third of hospital cases, and more than a third of the total deaths. The zymotic diseases, tuberculosis, much of the pneumonia and much of the heart disease, are infective in origin. Here, then, we have four types of disease—respiratory disease, the dyspepsias, general minor ailments, and the infective diseases—and to these we must add infant mortality. Now, a large proportion of these conditions are directly preventable, that is, we have the knowledge of causation and the potential power of prevention.

Further, it is known that the chief hindrance to the practice of prevention is lack of knowledge on the part of the public. Some illustrations may be named.

Through this one channel of infection it is now known that five principal diseases are conveyed, namely : pulmonary tuberculosis, pneumonia, influenza, poliomyelitis, and cerebro-spinal fever—diseases which in 1917 were responsible for one in five of all the deaths in England and Wales. These diseases are conveyed from person to person by the inhalation of the causal microbe. Protection cannot be secured by the application of sanitary measures to extraneous circumstances ; it is alone to be attained by methods of personal hygiene, applied on the individual scale of safeguarding one person from another. There is no other way. From dried expectora-tion, cough-spray, or the breath, the infection is conveyed. The infecting microbe may thus be carried directly from one person to another, or, as in pneumonia, it may be a normal inhabitant of the mouth, fauces, or nose which takes on a virulent form ; indeed, pneumonia is frequently an " auto-genic " infection, due to the invasion of the lung by pneumo-cocci present in the mouth of an apparently healthy person. Various conditions, such as cold, exposure, alcoholism, fatigue, or starvation, diminish the vitality and power of resistance of the lung and the pneumococcus obtains its foothold. Similarly, a lowered condition of vitality, or some acute disease such as measles or influenza, may call into activity the tubercle bacillus which might otherwise have lain quiescent in the body of the individual. From these facts it is obvious that serious disease and even death may result: (a) indirectly from dental caries, oral sepsis, sore throat, inflammation of the fauces, and similar conditions ; (b) directly from persons coughing or shouting at other persons or breath-ing in their faces. The proposition seems a very simple one, but the fact is that a clean mouth and clear respiratory passages, with a complete abstinence from spitting, and from sneezing, cough-spraying, shouting or breathing at other people, would go a long way towards the complete prevention of the five diseases named.

There is another large group of maladies, responsible for a high percentage of present physical impairment. The disabling indigestion and dyspepsia which brings patients in tens of thousands daily to the doctors and to the out-patient departments of the hospitals consists of indisposition which is at first of a functional nature. It is due partly to constitutional and atonic conditions of the alimentary tract and partly to mistakes in hygiene. " More than one-half of the chronic complaints which embitter the middle and latter part of life " said Sir Henry Thompson " is due to avoidable errors in diet ".

One cause or error is lack of knowledge in the choice of good and evil in regard to food—its selection, nutritive value, preparation, and consumption. But there are other aspects of the dietetic habit which have a large share in the production of disease : (a) irregularity of meals and the consumption of food or drink in between them ; (b) the unvaried monotony of the average dietary ; (c) the failure to masticate food owing to defective teeth or the habit of bolting food ; (d) drinking too much with meals ; (e) bad teeth ; (f) chronic constipation ; (g) swallowing air at the time of eating ; and (h) the abuse of tea, spices, and alcoholic beverages. These seem to be trivial matters, but it is out of the smaller habits and neglectfulness of men and women that disease is born. Again, there are the group of deficiency diseases : rickets, scurvy, and beri-beri, due in part to ignorance in dietetics. . . .

The story of the new knowledge with regard to infection constitutes one of the great romances of modern scientific achievement, a record which has made bright and immortal the times in which we live. It is a story, wonderful as regards attack, still more wonderful as regards defence, of those unseen friends and foes within the human body. It is a story of the infinitely little but with issues infinitely great, and which affect no less a question than man's life upon the earth. Always there are three factors in infection : the seed or infecting agent, the soil or body of the patient, and the sower or external circumstance which scatters the seed and prepares the soil— all those various external conditions which convey and influence the seed or affect the local or general vitality of the person (cold, fatigue, alcoholic excess, industrial surroundings, malnutrition, etc., and the age, sex, physique, and idiosyncrasy of the individual). In the middle of the sixteenth century Fracaster and Cardan first foretold the existence of germs of infection, and exactly 300 years later Louis Pasteur demonstrated their relation to disease. From 1870 to 1905 there followed the wonderful succession of discoveries which made known to us the causal microbes of the principal infective diseases. They may enter the body through four channels : by the alimentary canal through contaminated food, milk or water, as in typhoid or tubercle ; by the respiratory tract in influenza, diphtheria, pneumonia, and tubercle; by direct contact, as in smallpox, scarlet fever, or venereal disease ; or by inoculation through the skin by the bite of a louse or a mosquito.

One of the most devastating of the diseases of the European War was trench fever, at one time responsible for 60 per cent. of all cases of sickness. The British and American Trench Fever Commissions found that this mysterious malady, which disabled vast numbers of soldiers, is a specific and infectious disease caused by a virus in the blood, and that this virus is transmitted from man to man by the bite of the louse, and that the means of prevention is the avoidance of infestation with lice. Typhus is spread in Russia and Poland through the same medium of lice on the human body or clothing. Again, the bacillus of plague was discovered by Kitasato in 1894, and ten years later the Indian Plague Commission found that the ordinary mode of skin infection by this bacillus in man was by means of the bites of fleas, which had obtained the plague bacilli from rats. The grey rat of the sewers of the Oriental city abounds in fleas, and maintains the plague bacillus from season to season ; his cousin, the black house-rat, frequents the dwelling-houses, and is responsible for transmitting the disease to man. The death of rats from plague is thus the advent of its occurrence in man. But lice and fleas are not alone concerned in the spread of disease. There are sand flies, tsetse flies, and house flies. There is also the mosquito. Forty years ago Laveran discovered that malaria was due to a protozoon in the blood of the patient, and Manson and Ross showed that this parasite had a cycle of existence outside the human body in the mosquito, and that bites by a malaria-infected anopheles produced the disease in man. It is now known also that yellow fever is spread by another form of mosquito (stegomyia). Thus here we have lice, fleas, flies, and mosquitoes brought, in a handful of years, into a new, intimate, and profound relationship in our minds to the health of man. We may summarize the position in a sentence. The use of un-contaminated food, water, and milk ; breathing through the nose, and the avoidance of respiratory infection, of dried sputum and of cough spray ; the shunning of contagious diseases ; and no lice, no fleas, no rats, no mosquitoes : and we are on the high road to the abolition of some of the principal forms of human disease.

But the rest of the story is no less astonishing. The attack of these unseen foes in food, water, milk, dust, and in lice, fleas, and mosquitoes, is counteracted by the amazing resources of the human body—resources only now coming to the light of day. We now know that the defence of the body is fourfold. There is :

First, the resistant power arising from the physio-logical reserve of health, the power of hypertrophy and of increased functioning and metabolism in emergency.

Secondly, there is the control by the vasomotor nerve system which regulates the blood supply of any given part of the body, flushing it with the refreshing current of the blood stream.

Thirdly, there is a defence established by means of the

cells in the blood, the lymph and the tissues, which have the remarkable power of first catching and then absorbing into their own substance any invading germ or foreign element with which they come into contact.

And lastly, there is the newly discovered biochemical power represented partly by the normal secretions of the ductless glands (hormones), which stimulate into action the nervous system and the functions of various protective organs, and partly by the group of antitoxins which follow in the wake of toxins and are the direct reaction of the healthy body to their presence.

These four separate lines of defence are powerful, but it must be remembered that they never act in isolation. They are mutually interrelated, they co-operate together under a unified command, and they depend for their very existence upon a healthy and well-nourished state of the body. When, therefore, an infecting bacillus attacks man, it sets up, automatically, a chain of natural defences : (a) increased functional activity ; (b) a fuller blood supply ; (c) the stimulation of the catching and absorbing cells (phagocytosis) and the excitor secretions (hormones) ; and (d) a new formation of cells and substances antagonistic to, or assimilative of, the toxic products of the bacillus. When we ponder upon this array of defences called into operation by the act of infection, we cannot be surprised that one attack of a disease is not followed by another, and that a natural immunity against certain diseases is established. We begin to see the true philosophy of the action of antitoxins and vaccines ; we understand a little better the survival of man's body in Nature ; and we learn once and for all the necessity of bodily health as the strong and primary foundation of preventive medicine.

Sir George Newman has shown how a nation's health depends on the enlightened outlook of its citizens, but disease knows no frontiers, and if our defences against diseases are to be efficient they must be equally international. *How should a World Health Service be organized?* Dr. Vincent, who was President of the Rockefeller Foundation of Hygiene in New York from 1917 to 1929, shows how it ought to be done, in the following lecture. As a matter of fact some very considerable progress towards his ideal has been made—but on more democratic lines—as may have been gathered from Dr. Andrewes' remarks on the fight against influenza.

GEORGE EDGAR VINCENT

INTERNATIONAL PUBLIC HEALTH

IF the hygienic control of the world were put in the hands of some superman with scientific knowledge and authoritative powers, it is possible to imagine the way in which he would organize his forces and carry out his task. First of all, he would not rest content with the scientific resources in his possession, but would provide for continuous investigation in order to re-examine constantly the knowledge already acquired and to add new information. He would fit up well equipped and competently manned centres for investigation. These institutions he would place in strategic positions throughout the world in such a way as to bring the widest variety of diseases under constant scrutiny. He would further see to it that the staffs of these research centres were in constant communication, so that duplication of effort would be avoided and the results secured in one place put quickly at the disposal of workers in all the other institutions. In this way he would organize medical research as a world activity, constantly recruiting his groups of investigators and producing steadily new knowledge about the nature, cure, and especially the prevention of the maladies which afflict mankind.

In the second place, this hygienic force would effect an

administrative health organization throughout the world. Each country would be subdivided into sanitary districts, while the nations themselves would be brought into a unified administrative system under central control. Thus, stationed throughout the world, would be health officers with their technical experts and subordinates all organized into a hierarchy under single authoritative control. Under such a military régime sanitation, control of epidemics, regulation of individual conduct in the interests of health would be in force with all the efficiency which characterized the success of preventive measures in Cuba and the Panama Canal zone.

In the third place, through such an organization as has been described, there would be centralization of vital statistics—accurate and trustworthy conclusions based upon the reports of competent diagnosticians performing their duties in an objective way uninfluenced by economic and social considerations. These statistics, constantly gathered and interpreted by experts, would guide the organization and conduct of health campaigns in various parts of the world to meet situations as they were revealed by the statistical data. The appearance of an epidemic would be instantly reported to headquarters and orders would issue at once to apply measures which would ensure the prompt control of the threatened outbreak. Outposts and barriers against epidemics would be established and held in readiness for emergencies. Gradually the foci of diseases would be circumscribed until finally these sources of danger would be eliminated.

In the fourth place, the guardian of the world's health would not rest content with the negative control of disease. He would regard sanitation and epidemiology as only the first steps toward a positive campaign to be carried out through control of the diet, housing, exercise, and recreation of the world's population. This benevolent autocrat would see to it that children were well born and nourished and from their earliest days trained to hygienic habits. In this way he would, through the perfect obedience of millions to the dictation of wisdom and benevolence, produce a vigorous and healthy race.

Of course, as fundamental to this entire programme the commanding officer would establish medical schools and training centres in which the officers and privates of his army of hygiene would be recruited and trained for their important tasks. Under his guidance medical schools would transfer their chief interest from the cure to the prevention of diseases. The stress would be laid upon early diagnosis and upon preventive treatment, upon frequent medical examinations, upon sound habits of living, upon the importance of mental serenity and a stimulating social life. From such schools of medicine and hygiene would go out men and women apostles of the doctrine of prevention determined to keep people well and regarding the occurrence of disease among those under their charge as a serious reflection upon the vigilance and resourcefulness of the members of the profession.

The description of an arbitrary control of this kind is in itself sufficient to show how impossible and intolerable such a régime would be. Local autonomy, nationalistic feeling, the absence of supermen and human nature's resentment of control imposed from without are a few of the many obstacles which would make it utterly impossible to bring about the hygienic solidarity of the world. But the outlining of an imaginary unified system of control at least serves as a background against which to observe influences which have been going on for a long time in the world, but which of late have gained greatly in a definiteness of organization. In some sense, every one of the things which have been suggested as a part of an imaginary world organization is being to a degree accomplished.

Thus there are more than 445 medical schools scattered throughout the world. They represent the influence of three systems of medicine : the British, the Latin, and the German. These systems have been distributed in accordance with well-organized principles of national influence. The British system has prestige throughout the Empire, subject to modification especially in Canada. French medicine prevails in the Latin countries of Europe and in Central and South America; while the German plan is largely followed in Scandinavia, Austria, eastern Europe, in the Balkan regions, and in Japan. In the United States a combination of British and German methods has been gradually developed with certain features which are not found in either of the two European systems. The three national types are yielding to internal influences which may be counted upon to produce, not uniformity, but a more cosmopolitan ideal than at present exists anywhere in the world. In short, forces of intercourse and co-operation are approximating a world system more successfully than anything but the control of a superman could bring about.

Moreover, institutes for medical research are found in many parts of the world. Of these the best known are the Pasteur Institute of Paris, the Rockefeller Institute of New York, the London and Liverpool Schools of Tropical Medicine, the Hong Kong School of Tropical Medicine. Many other centres exist throughout the world. These centres, through publications, migrations, and international congresses, are kept in close communication, so that knowledge of each other's work is quickly disseminated. In a very true sense there is an informal and effective world organization for the prosecution of medical research. Many of the medical schools are an integral part of this system, making important contributions to what is a genuine international product.

In the collection of vital statistics definite progress towards world organization has been made. Through the International Office of Hygiene in Paris, forty nations are regularly exchanging information with respect to vital statistics. Uniform methods of reporting deaths have been agreed upon and are to a considerable degree being successfully carried out. The areas from which trustworthy and acceptable reports are made in various countries are being gradually extended. It is true that only a beginning has been made, but these beginnings have sketched a programme of international co-operation in the gathering and distribution of vital statistics which will inevitably be steadily elaborated during succeeding decades.

Growth of health organizations in different countries is constantly recorded. The leading nations of the world are giving increasing attention to health administration ; administrative areas are being organized and trained sanitarians put in charge. Not only is this national organization going forward, but the nations are drawing closer together in their co-operation for world health. The first European conference to consider health problems was held in 1851. Twelve nations were represented. Concerted measures against cholera, plague, and yellow fever were adopted. Thereafter at intervals of a few years other congresses were called to ensure better team-work in conformity with the rapidly advancing knowledge of preventive medicine. In 1902 an International Sanitary Bureau was established in Washington by the Pan-American Union. Finally, in 1908, a permanent International Office of Hygiene, which has already been mentioned, was established in Paris.

[Under the League of Nations (and again under U.N.O.) international co-operation has been achieved in the war against disease by such things as prompt notification of epidemics, a standardizing of vaccines and sera, international conferences of health officers, and control of the drug traffic.]

The fact that great epidemic diseases disregard national boundaries forces upon nations the adoption of co-operative measures against the ravages of these diseases. The first congress in 1851 was called to consider ways of dealing with cholera, plague, and yellow fever. The most striking illustration of national co-operation was afforded in 1920 and 1921 by the special commission against typhus in eastern Europe. This campaign was organized under the auspices of the League of Nations and had the direct financial support of fourteen Governments. A sanitary barrier was erected in eastern Europe and the march of the disease was halted. There is every reason to expect that co-operation of this sort will become more frequent and that diseases which heretofore have flourished because they encountered only unorganized resistance will now face the united front of a world sanitary army.

After public-health officers have done all that they can in the way of sanitation and control of contagious diseases, there remains a majority of diseases with which only the individual can deal. A large part of the problem of public health resolves itself into the question of personal hygiene. Only by changing the health habits of millions can the level of national and world efficiency be raised. In attempting this task certain traits of human nature must be reckoned with. All the resources of modern education, publicity, and suggestion must be employed. Especially must the habits of children be formed while they are still in a plastic stage. Thus, attempts in popular education in personal hygiene are being made in many countries and by various methods. The movement is still in the stage of experiment and demonstration. There are some signs of definite accomplishment in this field, but much remains to be done in the way of testing the material and methods of instruction before the campaign can be pushed with complete conviction and hope of permanent success. . . .

Thus as we try to get a picture of what is going on to-day in the world in the interests of the health of all the nations, we see that there is something like an approximation to the ideal system which a superman might conceivably try to set up. As compared with what remains to be accomplished, only an insignificant start has been made, but looked at from the standpoint of a century ago, really striking progress has already been achieved. In the last half-century the scientific resources of modern medicine have been enormously enriched. The causes of a great number of devastating diseases have been discovered ; the methods of controlling them have been worked out ; medical education has been put upon a higher level ; a beginning has been made in the training of expert sanitarians and an entire hygienic personnel. Organization of health administration has been greatly increased in efficiency. The death-rates in all the leading countries of the world have fallen in a most gratifying fashion. Beginnings have been made in the education of the public in the laws of personal hygiene. International understandings and good-will have been promoted. He would be hopeful indeed who should at the present time see anything like a millennium of human brotherhood ; but at any rate it is obvious that the tendencies now to be seen in the world toward co-operation for health cannot fail to draw scientific men everywhere into closer comradeship. So much is clear gain. There is reason to hope that for a time at least the resources of science will be turned from the destruction of human life to the healing of the nations.

Quite an appreciable reduction in the death-rate has been achieved in recent years, and much of the credit for this must be given to the new weapons against disease germs, the " antibiotics ", chief of which, of course, is penicillin.

Disease germs multiply by the simple process of splitting in two. Suppose a mere half-dozen germs gain entrance to the body through a cut in the skin, or through the soft tissues of the throat. Suppose they find conditions favourable, and they are able to double their numbers every four hours. It only needs three days for the half-dozen to become a million and a half ! No wonder the body's defences, well able to deal with hundreds of germs, or even with a larger number if given time, are completely overwhelmed by the sudden onslaught of these big battalions.

Now the ordinary antiseptics kill only those germs with which they come in contact, and they are unable to follow the germs into the blood-stream, or into the tissues of the body. Most of them indeed, are just as lethal to the cells of the body as they are to the germs. *What is the ideal antiseptic ?* What the doctors have been asking for is a harmless substance which can be carried to any part of the body in the blood-stream, and which will kill the germs, or at least prevent them multiplying, wherever they might be. Some of the steps towards the achievement of the ideal are described in the following extract from

the Linacre lecture of 1946, given by Sir Alexander Fleming, the
discoverer of penicillin. Since that time much further progress has
been made in our knowledge of the natural drugs produced by moulds
and other simple organisms, and two of them, chloromycetin and
aureomycin, promise to be most useful in dealing with diseases
caused by virus infection.

SIR ALEXANDER FLEMING

CHEMOTHERAPY

BACTERIAL infections have existed since time im-
memorial, and physicians, in all ages, have tried to deal
effectively with them. It has been our good fortune to have
lived in an era when many of these infections have, for the
first time, been brought under control, and there is promise
of more advances in the near future.

Until the middle of last century there was practically no
knowledge of the bacterial nature of the infections, so before
that time everyone was working in the dark, and many and
curious were the prescriptions used in the combat against
bacterial infections ; but we shall get no profit by discussing
these.

In the consideration of almost every branch of bacteriology
we go back to Pasteur and the latter half of the nineteenth
century. Pasteur proved that certain fermentations were due
to the action of microbes and that microbes were living objects
which did not arise *de novo* from the putrescible material
but were descendants of previously existing organisms.
Pasteur himself did not do any serious work on chemotherapy,
but his earlier bacteriological work stimulated Lister, who
had putrefaction of wounds very much at heart, to engage
himself on the subject.

There are some who use the word chemotherapy in a very
limited sense to cover those methods in which the chemical is
administered in such a way that it gets into the blood and

attacks the infecting microbes through the circulation in concentrations sufficient to destroy them or modify their growth. This is too narrow a definition and I shall use chemotherapy to cover any treatment in which a chemical is administered in a manner directly injurious to the microbes infecting the body. In this latter sense antiseptic treatment comes under chemotherapy—call it local chemotherapy if you like. The same general laws govern the treatment whether it is local or systemic, but there are certain particulars in which one has to draw a distinction. There are many chemicals used locally which are so poisonous to the human organism as a whole that they cannot be used for systemic treatment, but large numbers of these, although they had considerable vogue in the past when perhaps there was nothing better, are practically useless as chemotherapeutic agents except in the prophylactic sense. On the other hand, there are some chemicals, e.g. the sulphonamides, which are powerful agents for systemic treatment but which are frequently of little use when locally applied to a suppurating area, as their action is neutralized by substances occurring in the pus.

If a chemical is to be effective in the treatment of established infection it is necessary that, in addition to killing or inhibiting the growth of the microbes on the surface, it should be able to diffuse into the tissues to reach the microbes there. In a septic wound there are, of course, microbes in the cavity of the wound, but far more important are those which have invaded the walls.

Lister, like all other surgeons of ninety years ago, was struggling with the problem of septic wounds. To him, Pasteur's work came as a ray of light in the darkness. Putrefaction was due to living microbes which were introduced from outside. He set to work to prevent them being introduced. He cleansed his hands and his instruments, and treated them with chemicals, and he used a carbolic spray to kill bacteria in the air and prevent them reaching his operation wounds. In this way he revolutionized surgery.

[The antiseptics, however, could only kill the germs with which they came in contact in the wound. They could not attack those which had passed into the blood stream or the tissues. Indeed, in certain concentrations, every antiseptic did more harm than good by destroying the natural anti-bacterial power of the blood.]

Let us now come to chemotherapy in the narrower sense, *i.e.* the attack on the infective microbe through the circulation.

Scientific chemotherapy dates from Ehrlich and scientific chemotherapy of a bacterial disease from Ehrlich's Salvarsan, which in 1910 revolutionized the treatment of syphilis. The story of Salvarsan has often been told, and I need not go further into it except to say that it was the first real success in the chemotherapeutic treatment of a bacterial disease. Ehrlich originally aimed at *therapia magna sterilisans*, which can be explained as a blitz sufficient to destroy at once all the infecting microbes. This idea was not quite realized, and now the treatment of syphilis with arsenical preparations is a long-drawn-out affair. But it was extraordinarily successful treatment, and stimulated work on further chemotherapeutic drugs. While they had success in some parasitic diseases the ordinary bacteria which infect us were still unaffected.

It was in 1932 that a sulphonamide of the dye chrysoidine was prepared, and in 1935 Domagk showed that this compound (Prontosil) had a curative action on mice infected with strepto-cocci. It was only in 1936, however, that its extraordinary clinical action in streptococcal septicaemia in man was brought out. Thus twenty-six years after Ehrlich had made history by producing Salvarsan, the medical world woke up to find another drug which controlled a bacterial disease. Not a venereal disease this time, but a common septic infection which unfortunately not infrequently supervened in one of the necessary events of life—childbirth.

Before the announcement of the merits of the drug, Prontosil, the industrialists concerned had perfected their preparations and patents. Fortunately for the world, however, Tréfouel and his colleagues in Paris soon showed that Prontosil acted by being broken up in the body with the liberation of sulphanilamide, and this simple drug, on which there were no patents, would do all that Prontosil could do. Sulphanilamide affected streptococcal, gonococcal, and meningococcal infections as well as *B. coli* infections in the urinary tract, but it was too weak to deal with infections due to organisms like pneumococci and staphylococci.

Two years later Ewins produced sulphapyridine—another drug of the same series—and Whitby showed that this was powerful enough to deal with pneumococcal infections. This again created a great stir, for pneumonia is a condition which may come to every home.

The hunt was now on, and chemists everywhere were preparing new sulphonamides — sulphathiazole appeared, which was still more powerful on streptococci and pneumococci than its predecessors, and which could clinically affect generalized staphylococcal infections.

Fildes introduced a most attractive theory of the action of chemotherapeutic drugs. It was that these drugs had a chemical structure so similar to an " essential metabolite " of the sensitive organism that it deluded the organism into the belief that it was the essential metabolite. The organism therefore took it up, and then its receptors became filled with the drug so that it was unable to take up the essential metabolite which was necessary for its growth. Thus it was prevented from growing and died or was an easy prey for the body cells. This theory has been supported by many experimental facts and may give a most profitable guide to future advances in chemotherapy.

But another completely different type of chemotherapeutic drug appeared, namely, penicillin. This actually was described years before the sulphonamides appeared, but it was only concentrated sufficiently for practical chemotherapeutic use in 1940.

The story of penicillin has often been told in the last few years. How, in 1928, a mould spore contaminating one of my culture plates at St. Mary's Hospital produced an effect which called for investigation; how I found that this mould—a *Penicillium*—made in its growth a diffusible and very selective antibacterial agent which I christened Penicillin; how this substance, unlike the older antiseptics, killed the bacteria but was non-toxic to animals or to human leucocytes; how I failed to concentrate this substance from lack of sufficient chemical assistance, so that it was only ten years afterwards, when chemotherapy of septic infections was a predominant thought in the physician's mind, that Florey and his colleagues at Oxford embarked on a study of antibiotic substances, and succeeded in concentrating penicillin and showing its wonderful therapeutic properties; how this happened at a critical stage of the war, and how they took their information to America and induced the authorities there to produce penicillin on a large scale; how the Americans improved methods of production so that on D-Day there was enough penicillin for every wounded man who needed it, and how this result was obtained by the closest co-operation between Governments, industrialists, scientists, and workmen on both sides of the Atlantic without thought of patents or other restrictive measures. Everyone had a near relative in the fighting line and there was the urge to help him, so progress and production went on at an unprecedented pace.

Penicillin is the most powerful chemotherapeutic drug yet introduced. Even when it is diluted 80,000,000 times it will still inhibit the growth of *Staphylococcus*. This is a formidable dilution, but the figure conveys little except a series of many noughts. Suppose we translate it into something concrete. If a drop of water is diluted 80,000,000 times it would fill over 6000 whisky bottles.

We have already seen that all the older antiseptics were more toxic to leucocytes than to bacteria. The sulphonamides were much more toxic to bacteria than to leucocytes, but they had some poisonous action on the whole human organism. Here in penicillin we had a substance extremely toxic to some bacteria but almost completely non-toxic to man. And it not only stopped the growth of the bacteria, it killed them, so it was effective even if the natural protective mechanism of the body was deficient. It was effective too, in pus and in the presence of other substances which inhibited sulphonamide activity.

Penicillin has proved itself in war casualties and in a great variety of the ordinary civil illnesses, but it is specific, and there are many common infections on which it has no effect. Perhaps the most striking results have been in venereal disease. Gonococcal infections are eradicated with a single injection and syphilis in most cases by a treatment of under ten days. Subacute bacterial endocarditis, too, was a disease which, until recently, was almost invariably fatal. Now with penicillin treatment there are something like 70 per cent. recoveries.

It is to be hoped that penicillin will not be abused as were the sulphonamides. It is the only chemotherapeutic drug which has no toxic properties—in the ordinary sense of the word it is almost impossible to give an overdose—so there is no medical reason for underdosage. It is the administration of too small doses which leads to the production of resistant strains of bacteria, so the rule in penicillin treatment should be to give enough. If more than enough is given there is no harm to the patient but merely a little waste—but that is not serious when there is a plentiful supply.

But I am not giving you a discourse on penicillin. Suffice it to say that it has made medicine and surgery easier in many directions, and in the near future its merits will be proved in veterinary medicine and possibly in horticulture.

The spectacular success of penicillin has stimulated the most intensive research into other antibiotics in the hope of finding something as good or even better.

Waksman in 1943 described streptomycin, which is produced by *Streptomyces griseus*. This substance has very little toxicity and has a powerful action on many of the Gram-negative organisms. It has been used in tularæmia, undulant fever, typhoid fever, and *B. coli* infections, but the greatest interest has been in its action on the tubercle bacillus. *In vitro* it has a very powerful inhibitory action on this bacillus, and in guinea-pigs it has been shown to have a definite curative action. In man, however, the clinical results have not been entirely successful, but in streptomycin we have a chemical which does have *in vivo* a definite action on the tubercle bacillus, and which is relatively non-toxic. This is a great advance and may lead to startling results. One possible drawback may be that bacilli appear to acquire rapidly a fastness to streptomycin, much more rapidly than they do to penicillin or even the sulphonamides.

Many other antibiotics have been described in the last five years. Most of them are too toxic for use, but there are some which so far have promise in preliminary experiments. Whether they are going to be valuable chemotherapeutic agents belongs to the future.

If medical science keeps making these wonderful discoveries, it looks as though the killing diseases will gradually be conquered, and all men will be able to look forward to a " green old age "—barring accidents and warfare, of course. For most people the main drawback to old age is that there is not enough of it. *How can we ensure our fair share ?*

On this important matter we consult Sir Humphry Rolleston who was Physician to King George V. The first essential is to choose one's parents wisely, for longevity does run in families. As that is beyond our powers, we must make the most of the other good advice which he offers.

SIR HUMPHRY DAVY ROLLESTON

CONCERNING OLD AGE

IN considering old age, of which it is to be hoped we shall all have a happy experience, for there is a normal and cheerful as well as a morbid crabbed, and unhappy form, it is hardly possible to avoid a few words on the duration of life. In the simplest forms of animal life—the protozoa—which consist of a single cell, and so stand in the same relation to man that a brick does to a city, multiplication occurs by fission or division into two halves, and as there are now two cells instead of one and no vestige of a corpse, the organism is, as Weismann first insisted, immortal. This process may go on indefinitely, and has been watched by Woodruff in paramecium, during thirteen and a half years, for 8500 generations, comparable to a quarter of a million years of human life, without the occurrence of conjugation—namely, the union of two previously separate organisms, as occurs in the reproduction of higher animals—though periodically rejuvenation appears to take place by means of an independent internal reorganization (endomixis) of a single cell.

As the protozoa are, accidents apart, immortal, why is it that animals much higher in the scale never are ? It has been shown by culture experiments that the individual cells of man's body are also potentially immortal, but that the necessary conditions for this cannot be realized when they form part of a highly differentiated and specialized complex whole. In the higher walks of the animal kingdom rejuvenation of the constituent cells is thus rendered much more difficult or impossible, and the process of senescence must be regarded as the penalty for the high degree of individuation entailed in the complicated mechanism of the higher animals and man.

But it is interesting to recall that in the humble planarian flat worms the life cycle may be dramatically changed : by starvation they not only become smaller, but their structure becomes simpler, and their life cycle is reversed by the process

of de-differentiation, reduction, or involution. By alternate
starving and feeding, these flat worms have been kept stationary,
while others passed through nineteen generations, thus showing that the duration of existence of cells is not so much a
matter of time as of the work or metabolic changes that go
on inside them, although it may well be said that such a prolonged existence is not " life " according to some views ;
and although it is a far cry from amœbæ and flat worms to
man, these biological results may have some remote bearing
on the conditions influencing the duration of man's life. Not
only diet—over- and under-feeding, the effects of gross feeding
being obviously harmful—but the more mysterious influence
of the internal secretions, provided by the ductless or endocrine
glands, the thyroid, pituitary, and the gonads, the activities
of which may well be modified by diet, must be taken into
account.   Disturbance of the endocrine balance, or the
equilibrium normally maintained between the hormones or
chemical messengers of these endocrine glands has well-
recognized effects on growth and metabolism, and so may
influence the rate at which the body lives and wears or rusts
out. The significance of the much-discussed " rejuvenation "
produced by Steinach's and Voronoff's operations on the
sexual organs prove at any rate that there is much to learn,
especially whether or not the duration of life is really prolonged
thereby.

[Sir Humphry referred to the ages of the Biblical patriarchs,
and then gave some examples of " well-known if not well-
authenticated " centenarians of more recent days. He showed
how mistakes might easily have occurred by quoting the following from Raymond Pearl :

" John Smith was born in the latter half of the eighteenth century. His baptism was duly and properly
registered. He unfortunately died at the age of, say, 15.
By an oversight his death was not registered. In the
same year that he died, another male child was born

to the same parents, and given the name of John Smith, in commemoration, perhaps, of his deceased brother. This second John Smith was never baptized. He attained the age of 85 years, and then, because of the appearance of extreme senility which he presented, his stated age increased by leaps and bounds. A study of the baptismal records of the town disclosed the apparent fact that he was just 100 years old. The case goes out to the public as an unusually well-authenticated case of centenarianism, when of course it is nothing of the sort."]

FACTORS INFLUENCING LONGEVITY

Experience and that pure and reformed summary of it known as statistics show, as Karl Pearson and A. Graham Bell's figures prove, that inheritance is one of the strongest factors, if not the dominant one, in determining the span of life. The other factors are extrinsic and may be included under the heading of environment in its broadest sense, if we may now use a word which in 1835 John Sterling, in reproaching Carlyle for so doing, described as " barbarous " and " without authority ".

Numerous examples of families with long- or with short-lived tendencies will occur to everyone ; but what heredity can do is perhaps more convincingly established by Raymond Pearl's striking table, from Bell's analysis of the Hyde family, showing the influence of parental ages on that of the offspring : among 184 persons whose parents both lived more than 80 years the average age at death was 52·7 years, whereas among 128 persons whose parents died before 60 years of age their average life was 32·8 years, or nearly 20 years less. Forsyth's investigation, carried out on strict actuarial principles, shows that if the reasonably preventable diseases be eliminated the expectation of life, though increased, would not be so much prolonged as by the influence of a long-lived heritage. We should therefore, both in our own and in their interests, counsel our children to choose their parents carefully. As Benjamin Ward Richardson pointed out, the combined ages of one's parents and grandparents divided by six may be of assistance in numbering our days.

A sound stock may overcome the evil influence of environment, such as alcoholism and unhealthy surroundings in towns and thus explain the occasional longevity of those whose lives have been far from blameless, and the contrast between two aged brothers, one sober, the other intemperate. But even with a poor inheritance care may extend the term of a useful and happy life. Sir Hermann Weber, by practising the precepts of his book on *The Prolongation of Life*, led an active and happy existence until within a short time of his death at 95, although his parents died at or before 60 of heart failure and cerebral hæmorrhage—conditions due to high blood-pressure and arteriosclerosis (hardening of the arteries). . . .

Bodily Conformation.—The long-lived are usually spare and very seldom fat, neither very tall nor very short, very heavy nor very light. Insurance companies are naturally much interested in the bearing of the body-build on longevity, and in America much statistical investigation has been started into problems such as the bearing, in persons above the average weight, of the length of the spine, and a comparison of the chest and abdominal girths. From the insurance companies' statistics Dublin found the following rather surprising and confusing results : among those over the average weight the prospect of life was better among short and medium height men with short relative spine-lengths and with chest girths below the average, and among overweight tall men the outlook was best in those with long relative spine-lengths and chest girths below the average. So that in both of these groups of men above the average weight the presence of a chest measurement below the average was an asset.

Taking *environment* in its widest sense, the influence of disease in damaging the tissue cells and in accelerating degeneration and atrophic processes resembling those which normally occur must be given due weight. It has been said that only those who have kept their bodies free from disease up to the age of 60 can expect to attain extreme old age ; but nearly half of the 824 persons between the ages of 80 and 100 analysed by Sir George Humphry had passed through a severe illness, many of them acute infections. It seems highly probable, however, that the influence of an acute illness, the resulting changes being often transient and recoverable, would be much less harmful than that of a long-continued infection or intoxication in producing permanent changes. There are exceptions, and a terribly notable one is encephalitis epidemica, the " sleepy sickness " of the lay press. Chronic infections, especially those in connexion with the teeth, are prone to cause rheumatoid arthritis, and the resulting limitation of activity does harm both to the general health and mentally by impressing the idea of future crippling and incapacity. Acute illness may undoubtedly be the apparent starting-point of old age, more particularly if insufficient time, holiday, and change of scene be not allowed for recovery, the time allotted increasing roughly with the nature of the illness and the patient's age. The influence of infections, such as malaria, which Dr. W. H. S. Jones has shown was an important factor in the decadence of Magna Græcia, and of syphilis, both on the collective and individual health and longevity, need not be laboured.

The ideal of medicine is the prevention rather than the cure of disease, and for this end the detection of the earliest stages, and better of the disposing causes of diseases, is essential. Timely warning about diet, exercise, and manner of life may do much to prevent disease from getting a firm seat on a man's back, and it is not without significance that life assurance companies in America have found it pay to provide periodic medical overhauls to their clients.

Functional activity, mental and physical, plays a great part in keeping the body, when free from disease, trim and slim and

in postponing the advent of morbid old age. Occupation with
a keen desire to carry it through is so beneficial that some, such
as Karl Marx, would regard old age as largely a question of
will power. Speaking of the circle in which Madame du
Deffand moved, Lytton Strachey says :

> " They refused to grow old ; they almost refused to
> die. Time himself seems to have joined their circle,
> to have been infected with their politeness, and to have
> absolved them, to the furthest possible point, from the
> operation of his laws. Voltaire, d'Argental, Moncrif,
> Henault, Madame d'Egmont, Madame du Deffand
> herself, all lived to be well over eighty, with the full
> zest of their activities unimpaired."

Retirement is a problem beset with anxiety and danger ;
a successful business man when relieved of routine and able
to indulge in idle luxury and without hobbies may rapidly
degenerate. He now has to find occupation to kill time instead
of time to do all he must ; he begins to feel that " his day is
done ", that he has little but his dinner to look forward to, and
that at last he is old and a " has been ". Thus auto-suggestion,
even if not helped by suggestion from outside, hurries him on
the downward path ; any trifling ailment, such as rheumatism
or indigestion due to over-feeding, may arouse very hypo-
chondriacal anticipations, and observation of contemporaries
with premature senile changes may feed the flame of this
destructive auto-suggestion. There is, therefore, a basis for
the idea that senility is catching, and for seeking the companion-
ship of the young and thereby letting auto-suggestion work
in a constructive direction. A well-occupied mind, a happy
disposition that thinketh no evil, naturally smiles instead of
frowning on a stranger or a new idea, free from anger, hatred,
and jealousy, the vice that gives no pleasure to anyone, and an
attitude of charity in its original and best sense to all, tend to
prolong life and make it a happy, healthy prelude to crossing
the bar. In Sir James Crichton-Browne's words, " the best

SIR H. D. ROLLESTON

antidote against senile decay is an active interest in human affairs, and those keep young longest who love most ".

Food.—Experience, ancient and modern, both lay, such as is set forth in Cornaro's *Discourses on a Sober and Temperate Life* (1558), and medical, as contained in the works of the great George Cheyne (who at one time weighed 32 stone before he became a vegetarian), author of *An Essay of Health and Long Life* (1724), Metchnikoff, Henry Thompson the surgeon, and Hermann Weber the physician, agree on the importance of moderation in food. Most centenarians have been small eaters, especially of meat, and there is this advantage in poverty and living on Abernethy's earned sixpence a day, or its present-day equivalent. Excess of food, though more gradual in its evil influence, is more generally destructive than alcoholism, and there is much wisdom in Montaigne's dictum, " Man does not die, he kills himself ", and in the more dramatic proverb, " You dig your grave with your teeth ". As already mentioned, in some lowly forms of life, such as planarian worms, partial starvation leads to rejuvenescence and prolongs life, and there have not been wanting " cures " on such economical lines. A simple diet throughout life, and, after growth is completed, obedience to the rule of ceasing to eat before a feeling of repletion cries " Hold, enough " can be confidently recommended.

The influence of alcoholic drinks has been much discussed, and here, as in most problems, the influence of personal predilections may unconsciously tinge our conclusions. But there seems no doubt from collective investigations, such as Sir George Humphry inspired and insurance companies provide, that hard drinkers are exceptional among the long-lived, and that among the characteristics of those who attain great length of days temperance finds a place. But in spite of the opinion of some enthusiasts that the choice of total abstinence is the only sure path to longevity, the recent investigation, undertaken with all a modern statistician's precautions against fallacies, by Professor Raymond Pearl on over 5000 individuals at Baltimore gives the perhaps not unwelcome verdict that " a moderate use of alcohol does not tend to shorten life ". Such an opinion, based on sound data, is of unquestionable value ; and this I gladly admit, for my impression was that long-continued and constant absorption of alcoholic drinks, though always moderate and never amounting to intoxication, did tend to age the conscientious devotee ; but a few positive instances are apt to make an undue impression. What constitutes a " moderate use " is open to the criticism brought by Sir Hermann Weber, that what may be so for one is excess for another ; and the proverb, " Wine is the milk of old people " has been countered, as is the habit of proverb quoters, with " *Vinum lac veneris* " (wine is the milk of passion), and the danger of exposure to other risks. In forming a conclusion in any given case about the question, to drink, or not to drink, the personal equation of the individual must weigh heavily. No doubt total abstinence may suit some better and others worse, and while it certainly eliminates the problem of what is moderate, it equally deprives the aged of the help and comfort that some undoubtedly thus receive.

Smoking is on rather a different footing from excessive eating and drinking ; as Calverley said, " Stories, I know, are told not to thy credit ", but whatever may be true of the effects of the injection of nicotine into cats and rabbits, tobacco-smoking has not been proved to cause arteriosclerosis in man. As Sir Clifford Allbutt, a non-smoker and with a peculiar susceptibility to tobacco smoke, pointed out, if in any way it does cause arterial disease the process is so slow, at any rate in most people, that its effects become so mingled with other manifestations of old age as to be almost impossible to discriminate ; further, as Ruffer proved by examination of mummies, the ancient Egyptians (1580 B.C.-A.D. 525) certainly had arteriosclerosis without the consolation of tobacco. There is no doubt that the tolerance, painfully acquired in youth, to tobacco commonly diminishes with advancing years, and that unpleasant symptoms ranging from irregularity of the heart, abdominal pain, to tobacco angina may be the means of transforming a previously inveterate smoker into a total abstainer. Sir George Humphry, also a non-smoker, specially investigated the habits in this respect of centenarians; among 19 men 8 smoked much, 1 a little, 10 not at all; of 30 female centenarians 4 smoked much, 2 moderately, and 24 not at all.

There are statistics to show that people in the country live longer than those in towns.

The advice, then, to give others and even to practise ourselves should include a judicious choice of parents, avoidance of disease and worry, moderation in all things, mental and physical exercise, and open-air life, serenity and charity to all men. Leonard Williams's epigrammatic summary is easier to remember: " Fresh air, meagre food, freedom from care ".

The onset of what is popularly called old age varies in different countries: for example, the wheels of life run much faster in tropical countries, such as India, than in temperate latitudes. . . .

To put the beginning of old age at 50 or even 60 in man would no doubt raise the protest that it is only the elastic period of middle age. In truth there is so much variation that a rigid date cannot be fixed. In healthy people the advent of old age is so gradual that the individual himself has no suspicion of it, and very likely secretly preens himself on looking ten years younger than his years and his contemporaries, whose changed appearance arouses self-congratulation rather than self-examination. Perhaps the suspicion suddenly breaks upon him by overhearing a chance remark of others, by seeing an unwonted reflection of his figure in a mirror, or by some nice girl offering him her seat in an omnibus. Or a holiday may break to him that he cannot walk as of yore because of undue fatigue, or he is held up by shortness of breath or pain on unwonted exertion.

THE PHYSIOLOGY OF OLD AGE

While the whole body does not age at the same rate, there is a general and progressive diminution in functional activity corresponding to the atrophic involution of the cells of the organs and tissues. The response to stimuli of all sorts is diminished, and this sluggish reaction contrasts with the ready and comparatively exaggerated response to both normal and pathological stimuli in early life. The popular opinion that age is second childhood may be correct in that there is some resemblance between undeveloped and failing mental and physical powers, but there is an enormous difference between the ever actively moving child and the impassiveness of real old age. This failure of the power of reacting to stimuli is seen in the sense organs ; few people over 60 have perfect hearing, although most of us do not know it ; and apart from loss of acuity they often do not take in general conversation so well as others unless their attention is braced up for the purpose. The emotions are less active, the death of friends is less of a grief, and so the individual tends to become isolated and to live more on past than on present-day impressions ; hence a well-educated man may become more composed and satisfied, whereas one without intellectual interests may sink into mental torpor, vanity, and egocentricity, with the development of fads about health, undue garrulity, and a confirmed attitude of the *laudator temporis acti* (the praiser of times past). In what may be regarded as normal old age, psychical activity wanes ; new ideas do not bubble up, and when brought to his notice are not impetuously accepted. On the other hand, there is more of the philosophic calm born of what, when twitted with impatience, the young University don dismissed as " that greatly overrated property, experience ". Forgetfulness first of names, and much later of recent events, and mental fatigue are other evidences of the change. With loss of memory comes the habit of repeating the same story or remark, of mislaying things, and of becoming careless about the external graces. . . .

Muscular power and ability to walk as far and as fast as of yore diminish, the muscles and the glands of the intestine become sluggish so that constipation results, the diminished secretion of cutaneous perspiration makes the skin dry, and the bodily exchanges, as shown by basal metabolism estimations, are less. But except that the temperature in the axilla (armpit) is lower than normal on account of the diminished blood supply to the skin, the internal temperature of the body is not, as might be expected, and indeed has been stated, lowered. The explanation appears to be that although there is less heat produced in the body, there is less loss of heat from the dry skin on account of its poor vascular supply. Diminished sensibility to pain, both mental and bodily, is a beneficial relief, and suggests that with the gradual involution and approach to physiological death this warning will no longer be needed. This in some degree accounts for the latency of disease, such as pneumonia, urinary calculi, gallstones, or cancer, sometimes shown by the aged. There are, however, exceptions such as pain after zona (shingles) and obstinate itching. Sleep is less continuous than in youth, and early waking is common. Recovery from illness is slower, wounds and fractures of bones take longer to heal, and intracapsular fracture of the neck of the femur (thigh-bone), which is prone to occur from the thinning of the bone and the altered angle at the junction of the shaft and the neck of the bone, is a serious accident, often followed, as if it were the last straw, by dissolution. The old react somewhat differently to drugs, responding less promptly, so that some, such as purgatives, may be needed in considerable quantities. On the other hand, tolerance, as to tobacco, may become so much diminished that idiosyncrasies to certain drugs may appear ; morphine should be preceded by a previous dose of atropine to protect the senescent respiratory centre from being put to sleep, but the old are hardly so susceptible as children to this drug.

Though a gradually rising blood-pressure is commonly seen to accompany the passing years, a high blood-pressure is not a feature of old age ; in fact it might be said that those who attain great length of days do so, in part at least, because they have not had a high blood-pressure to wear out and thicken their blood-vessels. The pulse often shows inter-mittence (extra-systoles) ; this was so in one-fifth of the 824 cases analysed by Sir George Humphry, and though some-times a cause of annoyance, is not of any import. . . .

THE DESCRIPTION OF OLD AGE IN THE TWELFTH CHAPTER OF ECCLESIASTES

In connexion with old age everyone must recall the famous description in the first six verses of the twelfth chapter of Ecclesiastes, beginning, " Remember now thy Creator in the days of thy youth while the evil days come not, nor the years draw nigh when thou shalt say I have no pleasure in them ". Formerly ascribed to King Solomon (977 B.C.), the book of Ecclesiastes (in Hebrew, Koheleth = the preacher) has been shown by the higher criticism to date only from the end of the third century B.C. In his attractive work, *A Gentle Cynic*, the late Professor Morris Jastrow, jun., of Philadelphia, argued that the book of Ecclesiastes as it appears in the Authorized Version, consists of (i) the original, cynical, but good-natured *obiter dicta* of the unknown dilettante who preferred to veil his identity under the name of Koheleth ; and (ii) additions and modifications made by various hands to render it more orthodox and compatible with the tradition that it was written by Solomon. Thus the admonition " of making books there is no end, and much study is a weariness of the flesh " may very probably have been intended as a hint that Koheleth's views should not be taken too seriously. Following this con-ception, Jastrow reconstructed the text of the book of Ecclesi-astes to what he argued was its original form, and compared it with the more modern writings of Omar Khayyám and Hein-rich Heine. As we all must have speculated over the correct interpretation of the various metaphors in this description of the last stage of life, the explanations offered by others, such as Andreas Laurentius (1599), Master Peter Lowe (1612), founder of the Faculty of Physicians and Surgeons of Glasgow, Bishop J. Hall (1633), Richard Mead (1775), and Jastrow, may be very briefly mentioned. In 1666 John Smith devoted a book of 266 pages to elucidate these six verses which contain 207 words : *King Solomon's Portraiture of Old Age wherein is contained A Sacred Anatomy both of Soul and Body*. He is peculiar in authorship on this subject at his early age, 35 years, for nearly all the other writers on this topic, such as Cornaro, Sir Anthony Carlisle, Charcot, Sir George Humphry, Sir Hermann Weber, and Robert Saundby, not to mention and embarrass those happily with us, have been approaching " the sere, the yellow leaf ", and have perhaps been moved to their labours by the maxim, " Physician, cure thyself ".

The second verse, " While the sun, or the light, or the moon, or the stars, be not darkened, nor the clouds return after the rain ", is regarded by Laurentius, Lowe, and Hall as refer-ring to the ocular disabilities of old age, whereas Smith and Mead consider that mental failure and depression are meant. As regards the third verse, " In the day when the keepers of the house (the hands) shall tremble, and the strong men (the legs) shall bow themselves (become bent), and the grinders (teeth) cease because they are few, and those that look out of the windows (the eyes) be darkened ", there is general agree-ment, Lowe specially designating cataract as meant in the last sentence. " And the doors shall be shut in the streets " is regarded as referring to the mouth by Laurentius and Mead, and to the various orifices, including the results—constipation and dysuria—by Smith ; " when the sound of grinding is low " is considered by Jastrow to mean impaired hearing, and Smith as a lowered rate of metabolic processes, such as assimila-tion, blood formation, and various secretions. " And he shall rise up at the voice of the bird " implies, according to Smith and Mead, the early waking of the elderly ; " and all the daughters of music shall be brought low " signifies to Laurentius

the failure of voice, to Mead deafness, and to Smith all the
organs concerned with sounds—namely, the lips, tongue,
larynx, and the auditory apparatus. " Also when they shall
be afraid of that which is high, and fears shall be in the way "
is regarded by Smith as describing the general mental attitude
of anxiety for things both small and great and a bad head
for height ; but a more modern commentator suggests that
" afraid of that which is high " refers to dyspnœa (breathless-
ness) on climbing a hill. " And the almond tree shall flourish "
is by Laurentius, Hall, and Smith thought to refer to the white
hair or " churchyard flowers " of the old, but Mead argued
that loss of smell is meant. " And the grasshopper shall be
a burden " has been very variously interpreted : Hall is content to accept the literal meaning that the least weight is a
nuisance ; Laurentius and Lowe understand œdema (watery
swelling) of the legs ; John Smith that the aged body undergoes the reverse change of shrivelling, hardening, and angu-
larity. In the sixth verse the words, " Or ever the silver cord
be loosed ", refer, according to Laurentius, Lowe, Mead, and
Jastrow, to kyphosis (curvature of the spine), but Smith trans-
lates them into paralysis of the spinal cord and nerves. " Or
the golden bowl be broken " signifies cardiac failure to
Laurentius and Lowe, but cerebral hæmorrhage to Smith,
who thus explains the next line, " or the pitcher (the veins)
be broken at the fountain (the right ventricle), or the wheel
(the arterial circulation) broken at the cistern " (the left ventricle), and therefore concludes that King Solomon was perfectly acquainted with the circulation of the blood discovered
by William Harvey in 1616.

All this progress in medical science is very gratifying, but there is
one complication which will have to be faced, and this is the fact that
while the human population is increasing so rapidly the size of the
earth remains the same. The catastrophe which must inevitably

follow a steady increase in population has long been foreseen. It inspired Malthus's essay on population growth, which in turn gave Darwin the idea for his theory of evolution based on the struggle for existence. And even before Malthus we find Swift giving it for his opinion that " he who made two ears of corn or two blades of grass to grow upon a spot of ground where only one grew before would deserve better of mankind and do more essential service to his country than the whole race of politicians put together ".

It is, however, only in our own day that the problem of feeding the world's increasing millions has become a matter of urgency. In many countries of the world a steady decrease in the fertility of the soil has long been in progress ; the soil has been " exploited " rather than " farmed ". And soil erosion is fast reducing the amount of agricultural land available. It was said in 1949 that South Africa alone was losing 30 million tons of fertile top soil each year. Set against these losses an example of population increase : India, which can feed only 300 million people, already has over 400 millions, and every day some 14,000 new-comers arrive. Sanitation and the medical services have abolished the natural checks on population increase, so it cannot be wondered at that some people view the future with the utmost alarm, and begin to talk of " the dangerous doctor ". *Will biological progress prevent world famine ?* Sir James Scott Watson, who was Chief Scientific and Agricultural Adviser to the Ministry of Agriculture, reviews progress made in Britain in the past century, and finds grounds for optimism in the world-outlook.

SIR JAMES SCOTT WATSON

PRESSURE OF POPULATION ON LAND RESOURCES

YOU know one of our grave anxieties, as we look to the future, is about food supplies for the world's growing populations. As Mr. Dodd, the Director of Food and Agriculture Organization, has said : " Any night as we go to bed, we may ask ourselves what we can find for breakfast, for the 50,000 extra who will be with us in the morning ".

Let us look squarely at the facts, gloomy as they are. The

first is that very little is left of the once vast reserves of fertile
virgin land in the new world. Most of the possible farmland
that remains is, in one way or another, " problem country "—
swamp and semi-desert, dripping unhealthy tropical forest,
tundra and rocky hillside. Secondly, the world is constantly
losing farmland, partly by erosion and partly to provide for
needs other than food. And thirdly, the existing world population is very far from being adequately fed. The over-all
situation is worse than it was before the war, for although the
world is producing about 4 per cent. more food, there are
12 per cent. more people. We in this country have, of
course, cause for special concern. We are, it is true, better
endowed with land than some other nations—Japan, for
example ; but we can support our existing numbers, at our
present standard of living, only by importing more food than
any other country in the world.

Let us look back over some of the road that we have
travelled ; and let us start at 1851 when, to quote G. M.
Trevelyan : " The Great Exhibition spread its hospitable
glass roof high over the elm trees of Hyde Park, and all the
world came to admire England's wealth, progress and en-
lightenment ". Britain then had just ceased to be a pre-
dominantly rural country : the census of 1851 showed, for the
first time, that the majority of our people—though still only a
slender majority—were town-dwellers. But the responsibility for feeding our people still rested mainly on our home
farmers, for no overseas country was yet able to send us regular
or substantial quantities of grain ; and imports of meat, dairy
produce, fruit and other foods were negligible. Our farmers'
task had been constantly growing heavier. The population
of Great Britain and Ireland, though just about half what it is
to-day, had doubled since the first census was taken in 1801—
there were twice as many mouths to feed.

This growth of population had, of course, called for a great
and sustained effort on the part of our food producers—aimed
partly at higher yields, but more especially at adding acres to

the national farm. The sandy wastes of East Anglia's breck-land, the ragged commons of the midlands, the wide expanses of Scottish moors, Lincoln heath and Yorkshire wolds, as well as many little patches on the Welsh hillsides and along the fringes of Exmoor, had been converted, with much sweat, into cornfields and green pastures. Only one considerable block of the once vast expanse of the marshy, ague-ridden fens still remained to be drained. The general pattern of the country—except, of course, that towns were smaller and slag heaps fewer—was already much as we know it to-day.

But by the 'forties—the hungry 'forties—most of our fore-fathers knew in their hearts that we must give up the notion that we could continue to be self-supporting. The crisis had come in 1846 and 1847. By that time the potato had become an important crop—because it produced more food, per acre, than corn—and in Ireland it had replaced grain as the staff of life ; and in those two years the crop was stricken by a frightful new disease. The result was that 500,000 Irish people died of famine and pestilence, and one million more fled the country. So the Corn Laws were repealed—with the object, of course, of offering the inducement of an open market to overseas growers. But the flood of imports, hoped for by the hungry poor and dreaded by landowners and farmers, was not destined to arrive for another generation. By 1851 our reserves of unfarmed land were virtually exhausted and our farmers were concentrating on higher—that is to say, more intensive—farming. One vast operation, the drainage of our great expanse of wet clay, was well in hand. Every winter thousands of workers were busy, up to their knees in mud, digging thousands of miles of trenches, in which were laid millions of cylindrical clay tiles. New fertilizers were coming into use—nitrate from Chile, sulphate of ammonia from the new gas works, guano from the Pacific islands, superphosphate made by treating fossil bones with vitriol. But what caused most astonishment to our overseas visitors at the Great Exhibition was the vast range and the fine workmanship

displayed in the collection of British farm implements. Those
who went on to Windsor, to the great " Royal " Show, were
equally impressed with the great array of cattle and sheep.

The following quarter-century is commonly spoken of as
the Golden Age of British farming. By 1870 the old craft,
with the new aids which science was providing, reached an
astonishing level of efficiency, so that farmers and landowners
from other countries came thronging to see and to learn. For
in fighting our own battle we had forged many weapons that
others were later to turn to good account. It was the middle
'seventies which saw the beginnings of the flood of imported
wheat that had for so long been hoped for by the poor and
feared by the landed interests.

The reaper—one necessity for the mass production of corn—
had been at work for many years. And now the new railways
and ocean-going steamers could bring corn from the far-
distant new lands to British ports. As is now reckoned, the
world's resources of farmland had been increased by more
than 40 per cent. Another fact is that a good deal of the new
land, reserves of which seemed virtually inexhaustible, was
being exploited rather than farmed, so that the price of
wheat fell to a level that had hardly been known for three
centuries. And this was not all. In the early 'nineties,
cargoes of frozen meat began to arrive from the other side of
the world.

While the other countries of western Europe all did
something to keep their farms in being and so preserve their
rural people, we, for half a century, did nothing. Cheap food
seemed to our townsfolk, both rich and poor, to be an unmixed
blessing. Low costs of living enabled factory wages to be
kept down, and low wages meant that our manufactured goods
could be offered, overseas, at very attractive prices.

Not all of our farmers suffered grievously. There was a
growing demand for milk and vegetables, and home-produced
meat commanded a better price than the imported frozen
article. Some corn farmers, if they took the resolution in

time—before they went bankrupt—were able to swing over to livestock, fruit or vegetables. But in some eastern districts the majority of farmers broke, and their land either fell derelict or was rented for a song by " incomers " from Scotland, Wales, or Devonshire, who lived frugally in the kitchens of the great farm-houses and worked all the hours that God sent. In the uplands the bracken and heather crept down the hillsides into the fields that had but lately been reclaimed, and the midland clays, that had been so laboriously laid dry, reverted to rushes and briars. Between 1871 and 1931 the number of paid farm workers fell by more than half, and what was worse, those who remained came to feel that their job no longer mattered.

Of course, it is easy to be wise so long after the event. From a narrow economic point of view, and in the short term, our national policy was perhaps right. There were indeed many who, on social grounds or for sentimental reasons, regretted the decay of the countryside. But I cannot think of anyone who pointed out that, even from the strictly business point of view, our policy would, in the long term, prove wrong. And yet the argument could, I believe, have been made pretty convincing. In our country, population had quadrupled in less than two centuries, and there was no real doubt about the cause. It was not that the birth-rate had risen. It was simply that the progress of medical science and the improvement of medical services had vastly increased the expectation of life of every child that was born ; and it seems to me that it should have been clear, even in the 'eighties, that the same causes were beginning to operate over the world as a whole— I mean that world population would rise just as ours had risen. But there is nothing to be gained by crying over our past mistakes. For the past dozen years our farmers have had every encouragement to make good the harm that was done, and to press on to higher production. The level already reached is 40 per cent. above that of the pre-war period Moreover, new possibilities are constantly being created by the discoveries of science and by the native ingenuity of our farmers and workers.

What in particular have we done in Britain? I could make out a very long list, but a few examples must suffice. Penicillin, apart from its great value in human medicine, is rapidly eliminating a disease of dairy cows—mastitis—that used to cost millions of gallons of milk each month. BHC and other new insecticides are revolutionizing the war against destructive insects. The new hormone weed-killers were largely developed in this country during the war. We are again, as in the 'seventies, making immense strides in farm machinery and exporting great numbers of fine tools. The world's beef industry is still founded on our British breeds of cattle. Many of the great *estancieros* and station owners of the New World have been at the great Perth cattle sales this month. Again, we have given great thought to the organization of agricultural research, the education of future farmers and workers, and the machinery for the rapid dissemination of technical knowledge. In doing these things, and many more, we—like our grandfathers—are contributing notably to the world's stock of scientific knowledge and farming skill.

Will all this, with all that is being done elsewhere, be enough? Are we, as some believe, making headway towards the world of plenty? Or are the pessimists right? Will population outpace agricultural progress so that, in the end, the choice will lie between wars of extermination and universal famine?

It is true that there are physical limitations to the amount of food that can be produced on this planet by what you may call conventional methods. It is, as I think, conceivable that the world's soils could produce twice or even perhaps three times as much as they are doing to-day. But if population continues to double and redouble at intervals of seventy years—which is implied by the current statistics—what will happen in the coming centuries? Well, don't you think that posterity may decide that 4,000,000,000 or 5,000,000,000 are about as many people as this planet can comfortably accommodate ? And don't you think it conceivable that world population, like so many other things, could be planned ?

If not, there is still another possibility. You doubtless know that the basic process underlying all food production, at present, is the synthesis of sugar from air and water, which takes place in the leaf of the green plant exposed to the light of the sun. This same synthesis can be carried out in the laboratory. True, much remains to be done before the process could be carried out on a large scale ; and its operation upon a worth-while scale would require a vast expenditure of power. But isn't atomic energy in sight ?

I am not supposed to prophesy in this talk. But if I had to set up in that line of business, I should not yet be among the prophets of world famine.

The above lecture has led our thoughts into the future, and on this note our study of man, as biological science sees him, comes to an end. Progress has certainly occurred in the past—at least as regards certain adaptable species in every great class of plant or animal life. *Will evolutionary progress continue in the future ?* Will the poet's vision be realized :

> " These things shall be : a loftier race
>   Than e'er the world hath known shall rise,
> With flame of freedom in their souls
>   And light of knowledge in their eyes."

An answer to this vital question of the future of mankind is given in the following conclusion to Dr. Julian Huxley's book, *Evolution : The Modern Synthesis*. He points out that mankind has arisen through the emergence of " conscious and conceptual thought " and that our future evolution will not follow the same pattern as that of the lower animals in past ages. There will be no " adaptive radiation "—the evolutionary equivalent of the division of labour whereby any great step forward is followed by the gradual emergence of a large number of new species, each one of which exploits some particular aspect of the

new " invention ". Man must sink or swim as a single species : the
" unity of mankind " is certainly a biological fact of the most vital
significance.

The conclusion of the matter is that man's fate depends on man.
He has by now " become the trustee of evolution "—an immense
responsibility of which he hardly yet seems to be aware.

SIR JULIAN HUXLEY

PROGRESS IN THE EVOLUTIONARY
FUTURE

WHAT of the future ? In the past, every major step in
evolutionary progress has been followed by an outburst of change. For one thing the familiar possibilities of
adaptive radiation may be exploited anew by a number of
fresh types which dominate or extinguish the older dis-
pensation by the aid of the new piece of organic machinery
which they possess. Or, when the progressive step has opened
up new environmental realms, as was the case with lungs and
the shelled egg, these are conquered and peopled ; or the
fundamental progressive mechanism itself may be improved,
as was the case with temperature-regulation or the pre-natal
care of the young in mammals.

Conscious and conceptual thought is the latest step in
life's progress. It is, in the perspective of evolution, a very
recent one, having been taken perhaps only one or two and
certainly less than ten million years ago. Although already
it has been the cause of many and radical changes, its main
effects are indubitably still to come. What will they be ?
Prophetic fantasy is a dangerous pastime for a scientist, and
I do not propose to indulge it here. But at least we can exclude
certain possibilities. Man, we can be certain, is not within
any near future destined to break up into separate radiating
lines. For the first time in evolution a new major step in
biological progress will produce but a single species. The

genetic variety achieved elsewhere by radiating divergence will with us depend primarily upon crossing and recombination.

We can also set limits to the extension of his range. For the planet which he inhabits is limited, and adventures to other planets or other stars are possibilities for the remote future only.

During historic times, all or almost all of the increases in man's control over nature have been non-genetic, owing to his exploitation of his biologically unique capacity for tradition, whereby he is provided with a modificational substitute for genetic change. The realization of the possibilities thus available will continue to play a major part in human evolution for a very long period, and may contribute largely to human progress.

More basic, however, though much slower in operation, are changes in the genetic constitution of the species, and it is evident that the main part of any large genetic change in the biologically near future must then be sought in the improvement of the fundamental basis of human dominance—the feeling, thinking brain, and the most important aspect of such advance will be increased intelligence, which, as A. Huxley has stressed, implies greater disinterestedness and fuller control of emotional impulse.

First, let us remind ourselves that, as we have already set forth, we with our human type of society must give up any hope of developing such altruistic instincts as those of the social insects. It would be more correct to say that this is impossible so long as our species continues in its present reproductive habits. If we were to adopt the system advocated by Muller and Brewer, of separating the two functions of sex—love and reproduction—and using the gametes of a few highly endowed males to sire all the next generation, or if we could discover how to implement the suggestion of Haldane in his *Daedalus* and reproduce our species solely from selected germinal tissue-cultures, then all kinds of new possibilities would emerge. True castes might be developed, and some at least of them might be endowed with altruistic and communal impulses. In any case, as A. Huxley points out in an interesting discussion, progress (or, I would prefer to say, future human progress) is dependent on an increase of intra-specific co-operation until it preponderates over intraspecific competition.

Meanwhile there are many obvious ways in which the brain's level of performance could be genetically raised—in acuteness of perception, memory, synthetic grasp and intuition, analytic capacity, mental energy, creative power, balance and judgment. If for all these attributes of mind the average of our population could be raised to the level now attained by the best endowed ten-thousandth or even thousandth, that alone would be of far-reaching evolutionary significance. Nor is there any reason to suppose that such quantitative increase could not be pushed beyond its present upper limits.

Further, there are other faculties, the bare existence of which is as yet scarcely established : and these too might be developed until they were as commonly distributed as, say, musical or mathematical gifts are to-day. I refer to telepathy and other extra-sensory activities of mind, which the pains-taking work of Rhine, Tyrrell, and others is now forcing upon the scientific world as a subject demanding close analysis.

If this were so, it would be in a sense only a continuation of a process that has already been at work—the utilization by man for his own ends of hitherto useless by-products of his mental constitution. The earlier members of the Hominidæ can have had little use for the higher ranges of æsthetic creation or appreciation, for mathematics or pure intellectual construction. Yet to-day these play a large part in human existence, and have come to possess important practical consequences as well as value in and for themselves. The development of telepathic knowledge or feeling, if it really exists, would have equally important consequences, practical as well as intrinsic.

In any case, one important point should be borne in mind.

After most of the major progressive steps taken by life in the past, the progressive stock has found itself handicapped by characteristics developed in earlier phases, and has been forced to modify or abandon these to realize the full possibilities of the new phase. This evolutionary fact is perhaps most obvious in relation to the vertebrates' emergence from water on to land. But it applies in other cases too. The homothermy of mammals demanded the scrapping of scales and the substitution of hair ; man's erect posture brought with it a number of anatomical inconveniences. But man's step to conscious thought is perhaps more radical in this respect than any other.

By means of this new gift man has discovered how to grow food instead of hunting it, and to substitute extraneous sources of power for that derived from his own muscles. And for the satisfaction of a few instincts, he has been able to substitute new and more complex satisfactions, in the realm of morality, pure intellect, æsthetics, and creative activity.

The problem immediately poses itself whether man's muscular power and urges to hunting prowess may not often be a handicap to his new modes of control over his environment, and whether some of his inherited impulses and his simpler irrational satisfactions may not stand in the way of higher values and fuller enjoyment. The poet spoke of letting ape and tiger die. To this pair, the cynic later added the donkey, as more pervasive and in the long run more dangerous. The evolutionary biologist is tempted to ask whether the aim should not be to let the mammal die within us, so as the more effectually to permit the man to live.

Here the problems of values must be faced. Man differs from any previous dominant type in that he can consciously formulate values. And the realization of these, in relation to the priority determined by whatever scale of values is adopted, must accordingly be added to the criteria of biological progress, once advance has reached the human level. Furthermore, the introduction of such criteria based upon values, in addition

to the simpler and more objective criteria of increasing
control and the independence which sufficed for pre-human
evolution, alters the direction of progress. It might perhaps
be preferable to say that it alters the level on which progress
occurs. True human progress consists in increases of æsthetic,
intellectual, and spiritual experience and satisfaction.

Of course, increase of control and of independence is
necessary for the increase of these spiritual satisfactions ;
but the more or less measurable and objective control over
and independence of external environments are now merely
subsidiary mechanisms serving as the material basis for the
human type of progress ; and the really significant control
and independence apply to man's mental states—his control
of ideas to give intellectual satisfaction, of form and colour
or of sound to give æsthetic satisfaction, his independence of
inessential stimuli and ideas to give the satisfaction of mystic
detachment and inner ecstasy.

The ordinary man, or at least the ordinary poet, philosopher,
and theologian, is always asking himself what is the purpose
of human life, and is anxious to discover some extraneous
purpose to which he and humanity may conform. Some find
such a purpose exhibited directly in revealed religion ; others
think that they can uncover it from the facts of nature. One
of the commonest methods of this form of natural religion is to
point to evolution as manifesting such a purpose. The history
of life, it is asserted, manifests guidance on the part of some
external power ; and the usual deduction is that we can safely
trust that same power for further guidance in the future.

I believe this reasoning to be wholly false. The purpose
manifested in evolution, whether in adaptation, specialization,
or biological progress, is only an apparent purpose. It is as
much a product of blind forces as is the falling of a stone to
earth or the ebb and flow of the tides. It is we who have read
purpose into evolution, as earlier men projected will and
emotion into inorganic phenomena like storm or earthquake.
If we wish to work towards a purpose for the future of man, we

must formulate that purpose ourselves.   Purposes in life are made, not found.

But if we cannot discover a purpose in evolution, we can discern a direction—the line of evolutionary progress.   And this past direction can serve as a guide in formulating our purpose for the future.   Increase of control, increase of independence, increase of internal co-ordination ; increase of knowledge, of means for co-ordinating knowledge, of elaborateness and intensity of feeling—those are trends of the most general order.   If we do not continue them in the future, we cannot hope that we are in the main line of evolutionary progress any more than could a sea-urchin or a tapeworm.

As further advice to be gleaned from evolution there is the fact we have just discussed, that each major step in progress necessitates scrapping some of the achievements of previous advances.   But this warning remains as general as the positive guidance.   The precise formulation of human purpose cannot be decided on the basis of the past.   Each step in evolutionary progress has brought new problems, which have had to be solved on their own merits ; and with the new predominance of mind that has come with man, life finds its new problems even more unfamiliar than usual.   This last step marks a critical point in evolution, and has brought life into situations that differ in quality from those to which it was earlier accustomed.

The future of progressive evolution is the future of man. The future of man, if it is to be progress and not merely a standstill or a degeneration, must be guided by a deliberate purpose.   And this human purpose can only be formulated in terms of the new attributes achieved by life in becoming human.   Man, as we have stressed, is in many respects unique among animals : his purpose must take account of his unique features as well as of those he shares with other life.

Human purpose and the progress based upon it must accordingly be formulated in terms of human values ; but it must also take account of human needs and limitations, whether these be of a biological order, such as our dietary requirements or our mode of reproduction, or of a human order, such as our intellectual limitations or our inevitable subjection to emotional conflict.

Obviously the formulation of an agreed purpose for man as a whole will not be easy. There have been many attempts already. To-day we are experiencing the struggle between two opposed ideals—that of the subordination of the individual to the community, and that of his intrinsic superiority. Another struggle still in progress is between the idea of a purpose directed to a future life in a supernatural world, and one directed to progress in this existing world. Until such major conflicts are resolved, humanity can have no single major purpose, and progress can be but fitful and slow. Before progress can begin to be rapid, man must cease being afraid of his uniqueness, and must not continue to put off the responsibilities that are really his on to the shoulders of mythical gods or metaphysical absolutes.

But let us not forget that it is possible for progress to be achieved. After the disillusionment of the early twentieth century it has become as fashionable to deny the existence of progress and to brand the idea of it as a human illusion, as it was fashionable in the optimism of the nineteenth century to proclaim not only its existence but its inevitability. The truth is between the two extremes. Progress is a major fact of past evolution ; but it is limited to a few selected stocks. It may continue in the future, but it is not inevitable ; man, by now become the trustee of evolution, must work and plan if he is to achieve further progress for himself and so or life.

This limited and contingent progress is very different from the *deus ex machina* of nineteenth-century thought, and our optimism may well be tempered by reflection on the difficulties to be overcome. None the less, the demonstration of the existence of a general trend which can legitimately be called progress, and the definition of its limitations, will remain as a fundamental contribution of evolutionary biology to human thought.

All through this volume, as we said at the beginning, we have looked at life, low forms and high, from the admittedly narrow viewpoint of the scientist. But this does not mean that the scientist is unaware of the greater problems which lie beyond his present scope. *What are the greater problems of biology ?*

Some of these are eloquently stated below by Sir D'Arcy Thompson who was for many years, as *The Times* said, " one of the most respected and beloved figures in British university life ". The biologist can state these problems, but the answering must be left to the psychologist and the philosopher.

SIR D'ARCY WENTWORTH THOMPSON

THE GREATER PROBLEMS OF BIOLOGY

I LOVE to think of the logarithmic spiral that is engraven over the grave of that great anatomist, John Goodsir (as it was over that of the greatest of the Bernoullis), so graven because it interprets the form of every molluscan shell, of tusk and horn and claw, and many another organic form besides.

I like to dwell upon those lines of mechanical stress and strain in a bone that gave it its strength where strength is required, that Hermann Meyer and J. Wolff described, and on which Roux has bestowed some of his most thoughtful work ; or on the " streamlines " in the bodily form of fish or bird, from which the naval architect and the aviator have learned so much.

I admire that old paper of Peter Harting's in which he paved the way for investigation of the origin of spicules, and of all the questions of crystallization or pseudo-crystallization in presence of colloids, on which subject Lehmann has written his recent and beautiful book.

I sympathize with the efforts of Henking, Rhumbler, Hartog, Gallardo, Leduc, and others to explain on physical lines the phenomena of nuclear division. And, as I have said to-day, I believe that the forces of surface tension, elasticity, and pressure are adequate to account for a great multitude of the simpler phenomena, and the permutations and combinations thereof, that are illustrated in organic form.

I should gladly and easily have spent all my time this morning in dealing with these questions alone. But I was loath to do so, lest I should seem to overrate their importance, and to appear to you as an advocate of a purely mechanical biology.

I believe all these phenomena to have been unduly neglected and to call for more attention than they have received. But I know well that though we push such explanations to the uttermost, and learn much in so doing, they will not touch the hearts of the great problems that lie deeper than the physical plane. Over the ultimate organization of the living organism, we shall be left wondering still.

To a man of letters and the world like Addison, it came as a sort of revelation that light and colour were not objective things, but subjective, and that back of them lay only motion or vibration, some simple activity. And when he wrote his essay on these startling discoveries, he found for it, from Ovid, a motto well worth bearing in mind, *causa latet, vis est notissima* [the cause is hidden, the power is well known]. We may with advantage recollect it when we seek and find the force that produces a direct effect, but stand in utter perplexity before the manifold and transcendent meanings of the great word " cause ".

The similarity between organic forms and those that physical agencies are fit to produce, still leads some men, like Stephane Leduc, to doubt or to deny that there is any gulf between, and to hold that spontaneous generation or the artificial creating of the living is but a footstep away. Others, like Delage and many more, see in the contents of the cell only a complicated chemistry, and in variation only a change in the nature and arrangement of the chemical constituents ; they either cling to a belief in " heredity ", or (like Delage himself) replace it more or less completely by the effects of functional use and by chemical stimulation from without and from within. Yet others, like Felix Auerbach, still holding to a physical or quasi-physical theory of life, believe that in the living body the dissipation of energy is controlled by a guiding principle, as though by Clerk Maxwell's demons ; that for the living the law of entropy is thereby reversed ; and that life itself is that which has been evolved to counteract and battle with the dissipation of energy. Berthold, who first demonstrated the obedience to physical laws in the fundamental phenomena of the dividing cell or segmenting egg, recognizes, almost in the words of John Hunter, a quality in the living protoplasm, *sui generis*, whereby its maintenance, increase, and reproduction are achieved. Driesch, who began as a " mechanist ", now, as we have seen, harks back straight to Aristotle, to a twin or triple doctrine of the soul. And Bergson, rising into heights of metaphysics where the biologist *qua* biologist cannot climb, tells us (like Duran) that life transcends teleology, that the conceptions of mechanism and finality fail to satisfy, and that only " in the absolute do we live and move and have our being ". " We end but a little way from where we began."

With all the growth of knowledge, with all the help of all the sciences impinging on our own, it is yet manifest, I think, that the biologists of to-day are in no self-satisfied and exultant mood. The reasons and the reasoning that contented the past generation call for re-inquiry, and out of the old solutions new questions emerge ; and the ultimate problems are as inscrutable as of old. That which, above all things, we would explain, baffles explanation ; and that the living is a living organism tends to reassert itself as the biologist's fundamental conception and fact. Nor will even this concept serve us and suffice us when we approach the problems of consciousness and intelligence, and the mystery of the reasoning soul ; for these things are not for the biologist at all, but constitute the psychologist's scientific domain. In wonderment, says Aristotle, does philosophy begin, and more than once he rings the changes on that theme. Now, as in the beginning, wonderment and admiration are the portion of the biologist, as of all those who contemplate the heavens and the earth, the sea, and all that in them is.

And if wonderment springs, as again Aristotle tells us, from ignorance of the cause of things, it does not cease when we have traced and discovered the proximate causes, the physical causes, the efficient causes of our phenomena. For beyond and remote from physical causation lies the End, the Final Cause of the philosopher, the reason Why, in the which are hidden the problems of organic harmony and autonomy, and the mysteries of apparent purpose, adaptation, fitness, and design. Here, in the reign of teleology, the plain rationalism that guided us through the physical facts and causes begins to disappoint us, and intuition, which is of close kin to faith, begins to make herself heard.

And so it is that, as in wonderment does all philosophy begin, so in amazement does Plato tell us that all our philosophy comes to an end. Ever and anon, in the presence of *Magnalia Naturæ*, we feel inclined to say with the poet : " These things are not of to-day nor of yesterday, but evermore, and no man knoweth whence they came ".

I will not quote the noblest words of all that come into my mind, but only the lesser language of another of the greatest of the Greeks : " The ways of His thoughts are as paths in a wood thick with leaves and one seeth through them but a little way ".

GLOSSARY OF SCIENTIFIC TERMS

A. Ångstrom unit : One ten-millionth of a millimetre.

Afferent : Conducting inwards. Afferent nerves convey sensations to the brain. Opposite of efferent.

Alleles or Allelomorphs : The pair of corresponding genes which, like a pair of South African stamps in different languages, perform the same function, but give a different appearance.

Antibiotics : Lit. tending to prevent, injure, or destroy life. Hence the newly discovered substances like penicillin which prevent the multiplication of bacteria or viruses.

Arthropoda : The great phylum of animals with jointed limbs and segmented bodies ; including insects, spiders, shrimps, etc., and centipedes.

Bacteria : Vegetable micro-organisms of parasitic nature. Agents of putrefaction and fermentation, and the cause of many diseases. Grouped into spherical cocci, rod-shaped bacilli, and spiral spirilla. Many have cilia.

Birefringent : Splitting a ray of light into two rays which are bent out of their original path by different amounts.

Chondro- : Greek root meaning cartilage.

Chroma- : Greek root meaning colour.

Chromatin : The part of the nucleus of a cell which is easily coloured with dyes.

Chromosome : One of the bodies in the chromatin, like a string of beads, along which the genes are spaced.

Cilia : Hair-like processes on cells, capable of lashing movements.

Clone : A group of organisms which can be produced asexually or vegetatively, like the apple or potato, and which do not usually breed true from seed. The Greek root means a mob.

Cyto- : Greek root meaning a hollow vessel or cell.

Diploid : The double number of chromosomes found in all cells of the body except the germ cells. See Haploid.

Endo- : Greek root meaning within.

Enzyme : A chemical which effects the transformation of other compounds without itself changing.

Flagella : Long cilia. Usually one or two are sufficient to whip the organism along through the water or other liquid.

Gametes : The "marrying" cells : sperms, or eggs which unite with one of the other sex to form a new individual.

Genes : The units in the chromosomes which determine how the organism will resemble its parents.

Haploid : Having the halved number of chromosomes characteristic of the germ cells.

Hetero- : Greek root meaning other than, or different.

Holo- : Greek root meaning whole.

Homo- : Greek root meaning one and the same, common, or joint.

Homo : Latin for man.

Homothermy : Preserving a uniform body temperature ; warm-blooded.

Hominidæ : Man and the man-like extinct species.

Hominoidæ : The super-family which includes man and the apes.

Hyalo- : Greek root meaning glassy.

Hypertrophy : Abnormal growth or excessive development ; the opposite of atrophy.

Involution : Degeneration, or retrograde evolution.

Leucocyte : The white or colourless corpuscles of the blood. When they wander about in the soft tissues of the body, consuming degenerating tissues or disease germs, they are called phagocytes.

Meiosis : The reduction of the chromosome number to one-half the normal when the gametes (sperm or egg cells) are being formed.

Meta- : Greek prefix meaning a change of position or form or condition or time.

Metabolism : The sum total of chemical changes proceeding in a living cell or organism.

Mito- : Greek root meaning threadlike.

Mitosis : The process by which a cell-nucleus divides in two, with an exact division of the chromatin between the two daughter-cells.

Mutation : A sudden variation.

Neo- : Greek root meaning youthful or new.

Neoteinic : Having a prolonged period of youthful immaturity.

Osmosis : Diffusion through a membrane separating two liquids in different states. Usually tending to equalize conditions.

Parthenogenesis : Reproduction from eggs which develop without being fertilized, as happens with greenfly in summer.

Petiole : A leaf-stalk.

Plasmo- : Greek root meaning form ; hence the jelly-like or viscous material of a cell.

Pulvinus : The swelling on a petiole where it joins the main stem. The seat of irritability in leaves.

Polymerization : Changing to a substance of higher molecular weight, but retaining exactly the same proportions between the elements.

Protozoa : Animals or plants consisting of one cell only, but differing from bacteria by virtue of having a well-defined nucleus.

Soma- : Greek root meaning body. Hence a somatic cell is any cell of the body apart from the germ cells.

Staphylococcus : Any microbe which forms bunches or clusters.

Streptococcus : One of the cocci which form chains. The cause of many serious diseases.

Vasomotor : Controlling the size of the blood-vessels and hence the supply of blood.

Virus : Disease-producing agents on the border line between living and non-living. They can pass through filters which stop bacteria, and they also differ from bacteria in that they cannot be cultivated in cell-free material in the laboratory, but only in the living tissues of some host.

Zygote : The new individual formed by the union of two gametes.

BIOGRAPHICAL NOTES

MICHAEL ABERCROMBIE, F.R.S. (1912-   ), Reader in Embryology in the University of London, has carried out notable research work on the growth of new, and the regeneration of damaged, tissues. He has made a careful experimental analysis of the development of chick embryos, and a valuable study of the effects of injury on nerves and other organs. He was educated at Leeds Grammar School and Queen's College, Oxford, where he became Scholar, then Research Fellow, and was awarded the Beit Memorial Fellowship in Medical Research. After lecturing in Zoology at Birmingham he went to University College, London, as Lecturer in Embryology.

C. H. ANDREWES, F.R.S. (1896-   ), Deputy Director of the Scientific Staff of the National Institute for Medical Research since 1952, is also in charge of the World Influenza Centre. He started his medical education at St. Bartholomew's Hospital, served as a Surgeon Sub-Lieutenant in the R.N.V.R. in 1918-19, and then returned to complete his training, winning University gold medals for both surgery and medicine. He was House Physician at Bart's from 1921 to 1923 and again 1923-25, the interval being filled with a visit to America where he was Assistant Resident Physician in the Hospital of the Rockefeller Institute, New York. He has published a number of technical papers on viruses.

J. D. BERNAL, F.R.S. (1901-   ), has played a leading part in the development of the comparatively new field of biophysics. After graduating at Emmanuel College, Cambridge, he was a research worker at the Davy Faraday Laboratory from 1923 to 1927, and then Lecturer in Structural Crystallography at Cambridge until 1934, when he was appointed Assistant Director of Research in Crystallography. He was elected F.R.S. in 1937, became University Professor of Physics at Birkbeck College, London, in 1938, and won the Royal Medal of the Royal Society in 1945. His publications include: *The World, The Flesh and The Devil* ; *The Social Function of Science* ; *The Freedom of Necessity* ; *The Physical Basis of Life* ; and *Science in History*.

SIR JAGADIS CHANDRA BOSE (1856-1937), sometime Professor of Physical Science at Presidency College, Calcutta, and founder of the

Bose Research Institute, combined all that is best in Eastern and Western inspiration. After graduating in Calcutta he came to England for further study at Cambridge and in London. His genius as a research worker was widely recognized. He was knighted in 1917 and elected F.R.S. in 1920. An account of his main work is given in his book on *The Nervous Mechanism of Plants*.

ROBERT BURTON (1577-1640) spent most of his life at Oxford, and his famous book, *The Anatomy of Melancholy*, bears witness, in its multitude of quotations, to his quiet and studious habits. He graduated at Brasenose College, and was elected student of Christchurch in 1599. He held the livings of St. Thomas's, Oxford, and of Segrave in Leicestershire.

GUNNAR DAHLBERG (1893-) the world-famous Swedish geneticist, is Director of the State Institute of Human Genetics in the University of Uppsala. He studied medicine first, and then turned to the intensive consideration of the machinery of heredity, especially as applied to mankind. He has received a number of honorary degrees from British and other universities. His translated publications include *Twin Births* and *Twins from a Heredity Point of View* ; *Inbreeding in Man* ; and *Race, Reason and Rubbish*.

CHARLES DARWIN (1809-82) will always be remembered as the author of *The Origin of Species*, the book which first gave a convincing proof that evolution had occurred, and first suggested a likely process by which it had occurred. It was the evidence collected on the voyage of the *Beagle* which gave him the material for his theory, the first sketch of which was made in 1842, while a longer one was circulated among his friends in 1844. The book, however, was only published in 1859 after Darwin had received a manuscript from A. R. Wallace outlining the same ideas on which he had been working for so many years.

VISCOUNT DAWSON OF PENN (1869-1945) was the first practising physician to be raised to the peerage. He studied medicine at University College, London, and in the London Hospital, where he was appointed assistant physician in 1896 and physician in 1906. He specialized in gastric ailments and diabetes, and rapidly rose to the first rank, being appointed Physician-in-Ordinary to King George V (then Prince of Wales) in 1907. The appointment was confirmed at the King's accession and renewed by the two succeeding monarchs. Lord Dawson's many honours include a C.B. for war services, the

G.C.V.O., K.C.M.G., K.C.B., a peerage in 1920 and a viscountcy in 1936. He was President of the Royal College of Physicians (1931-38) and a member of the Medical Research Council.

ARTHUR DENDY, F.R.S. (1865-1925), was educated at Manchester and began his career in the Natural History section of the British Museum. He then sailed for Australia to become Demonstrator in Biology in the University of Melbourne, and after six years there he accepted the Chair of Biology at Canterbury College, University of New Zealand. In 1903 he became Professor of Zoology at Cape Town, and two years later returned to England to King's College in the University of London. Among his books are *Outlines of Evolutionary Biology* and *The Biological Foundations of Society*.

PHILIP EGGLETON (1903-54) was Reader in Biochemistry in the Department of Physiology at Edinburgh until his untimely death. He graduated in Chemistry at University College, London, in 1922, and turned at once to the application of his chemical knowledge to the problems of physiology. He worked for a time under Prof. A. V. Hill, and was awarded a Beit Memorial Research Fellowship in 1926. His researches threw new light on the mechanism of muscle contractions whereby we transform the energy of our food into work, and in 1931 he was awarded the Julius Mickle Fellowship for the best research carried out in London University during the previous five years. He went to Edinburgh in 1930 and did excellent work in both lecturing and research.

D. W. EWER (1913-   ), Professor of Zoology and Entomology in Rhodes University, was educated at Trinity College, Cambridge, and at Birmingham where he took his Ph.D. He has specialized in the study of lowly forms of life, the worms and insects, and that curious " missing link ", *Peripatus*, found in the tropics and in Australasia, which shows how the Arthropoda have probably developed from the segmented worms. His researches on physiology and anatomy were interrupted during the second world war while he attended to more urgent problems of anti-aircraft defence, and measures to deal with flying bombs and rockets.

SIR ALEXANDER FLEMING, F.R.S. (1881-1955), may have been lucky in his discovery of penicillin, but it was the sort of luck that comes to the man that deserves it. He was born in Ayrshire, and went to school at Kilmarnock. He studied medicine at St. Mary's in London, and carried off most of the class prizes and scholarships, taking honours in nearly every branch of medicine, and winning the University gold medal. He served as a Captain in the R.A.M.C. during the first world war and was mentioned in despatches. He then returned to his bacteriology lectures and his research at St. Mary's. It was while he was looking for the cause of influenza that the spores of the penicillin-producing mould settled on one of his culture plates and created a germ-free circle round about. Fleming was quick to see the importance of this chance-happening, and he devoted all his skill and energy to developing the discovery. The culture and isolation of penicillin proved to be a most difficult task, and that is why Sir Howard Florey and others shared the final triumph. Fleming, Florey and E. B. Chain were awarded the Nobel Prize for Medicine in 1945.

E. B. FORD, F.R.S. (1901-   ), formerly represented the Commonwealth on the Permanent International Committee of Genetics. He is Reader in Genetics and Director of the Genetics Laboratory at Oxford, and one of the founder-members of the Nature Conservancy. He graduated from Wadham College, Oxford, and was for some time University Lecturer and Demonstrator in Zoology and Comparative Anatomy. He was President of the Genetical Society of Great Britain from 1946 to 1949. His publications include: *Mendelism and Evolution*, a standard text-book; *The Study of Heredity*, in more popular style; *Genetics for Medical Students*; and a volume on *Butterflies* in the New Naturalist Series.

JOHN E. HARRIS, F.R.S. (1910-   ), Professor of Zoology in the University of Bristol since 1944, went from the City School, Lincoln, to Christ's College, Cambridge, where he started research work on the physical structure of protoplasm. He then turned to a study of the way in which fishes swim, and as a Commonwealth Fund Fellow he carried out further investigations on the hydrodynamics of fish locomotion at Columbia University from 1933 to 1935. He then became a Demonstrator in Zoology at Cambridge, and continued his work on marine biology with a Royal Society Messel Research Fellowship. During the war he directed a team of biologists and chemists working on the fouling of ships, becoming Chairman of the same committee after his appointment to Bristol. He is now returning to his early interest in cell physiology.

A. V. HILL, F.R.S., C.H. (1886-   ), was Foulerton Research Professor of the Royal Society from 1926 to 1951. He is a leading authority on the mechanism of muscle and nerve, and the chemical and physical changes accompanying all activity in those organs. His many honours include the Nobel Prize for Medicine in 1922, the O.B.E., the ribbon of the Companions of Honour, the American Medal of Freedom with Silver Palm, and the French Legion of Honour. Professor Hill has served on a large number of important scientific councils and committees. His popular books include *Living Machinery* and *Adventures in Biophysics*.

SIR JULIAN S. HUXLEY, F.R.S. (1887-   ), has passed from a study of evolution in the past to the quest of where it will lead in the future. Like his grandfather, T. H. Huxley, he has a clear and powerful mind which finds expression in a corresponding style. He went to Eton and Balliol College, Oxford, and held many posts, first as Lecturer and then as Professor of Zoology. He was Secretary of the London Zoological Society from 1935 to 1942, and a member of the committee which carried out Lord Hailey's African survey of 1933-38, while from 1946 to 1948 he was Director-General of U.N.E.S.C.O. He is a member of many scientific societies, and President of the Institute of Animal Behaviour. His numerous books include: *Africa View*; *Bird-watching and Bird-behaviour*; *At the Zoo*; *Evolution, the Modern Synthesis*; *Evolutionary Ethics*; and *Man in the Modern World*. He collaborated with Professor Andrade to write *An Introduction to Science*, and with H. G. Wells and G. P. Wells to write *The Science of Life*.

THOMAS HENRY HUXLEY (1825-95) will always be remembered for his vigorous and eloquent championship of the cause of evolution following the publication of Darwin's books. A great biologist himself, he fully appreciated the force of Darwin's arguments; a passionate seeker of truth, he was the avowed enemy of all prejudice and obscurantism; a firm believer in freedom of thought and speech, he worked hard to spread the light of knowledge. After studying medicine at Charing Cross Hospital he joined the navy, and made his first cruise in the *Rattlesnake*. His studies of marine plankton earned him a Fellowship of the Royal Society. In 1854 he became Professor of Natural History in the School of Mines. He spent much time in public lecturing and in writing essays, usually on controversial subjects, but always with unfailing courtesy and good temper.

LORD LISTER (1827-1912) performed the first antiseptic operation on August 12, 1865, a date which marks the dawn of modern surgery. Joseph Lister was born in Essex, the son of an F.R.S. to whom we owe the real effectiveness of the compound microscope. He was educated at Quaker schools and at University College, London, taking his B.A. in 1847 and then turning to medicine and surgery. From his earliest days as a surgeon he brooded over the terrible mystery of the blood poisoning, in various forms, which was such a scourge in the hospitals. He went to Edinburgh, where he became house surgeon to Syme, one of the greatest of all British surgeons, and in 1860 was appointed Professor of Surgery in Glasgow University. By this time he had arrived at the conclusion that cleanliness was the best defence against blood-poisoning, but the real nature of the cause of such poisoning eluded him until, in 1865, his attention was drawn to Pasteur's work on fermentation and putrefaction. This gave him the clue he needed. He started using carbolic acid as an antiseptic, first painting it on the surface of wounds, and then operating under a fine spray. Then he came to the conclusion that it was not air-borne germs that he had to fear so much as those on the hands, instruments, and dressings of the surgeon and his assistants. Thus aseptic surgery was born. Lister also discovered the value of the catgut ligature. He was raised to the peerage in 1897, and included in the first list of members of the Order of Merit in 1902.

WALTER ROBERT MATTHEWS, K.C.V.O. (1881-   ), the Dean of St. Paul's, has succeeded better than most of his colleagues in harmonizing the essential elements of the Christian faith with the discoveries of modern science. He has long been an ornament to both academic and clerical worlds. Educated at Camberwell and at King's College, London, he spent a few years in London curacies and then became Lecturer in Philosophy and in Theology at King's. From 1918 to 1932 he was Professor of the Philosophy of Religion, and also Dean of his College, and for much of that period Chaplain to the King. He became Dean of St. Paul's in 1934, and received the K.C.V.O. in 1935. Among his many books we might mention *Studies in Christian Philosophy*; *The Psychological Approach to Religion*; *God and Evolution*; *The Christian Faith*; and *The Foundations of Peace*.

J. G. MENDEL (1822-84) was born in Austrian Silesia, the son of a small peasant proprietor. He entered the monastery at Brünn (now Brno) at the age of 21, and after studying natural science at Vienna for three years he returned to Brünn as a teacher. His famous experiments on plants were carried out in the cloister garden, and his reports were published in the proceedings of the local Natural History Society, but they failed to attract more than purely local and temporary interest. He became abbot of his monastery in 1868, and then had little time for further researches, while such experiments as he did carry out served only to baffle and disappoint him owing to the complicated genetical pattern of the plants, like the hawkweeds, on which he worked. His reports were unearthed and their importance realized in 1900, when the championship of Professor William Bateson quickly spread them over the academic world.

LADY MARY WORTLEY MONTAGU (1689-1762) possessed a rare combination of beauty, wit, imagination, sound common sense, and wide learning. She was the daughter of the Duke of Kingston, who did not see what education could add to the blessing of high birth, but Lady Mary was wiser than her father, and she was encouraged in her self-imposed studies by her uncle and by Bishop Burnet. In 1712 she married Edward Montagu. It was a runaway match, and for a time the young couple lived in retirement in the country, but in 1715 Montagu became a Commissioner of the Treasury, and his wife took her rightful place among the wit and elegance of the court. A year later he was appointed Ambassador to Constantinople, and it was there that Lady Mary made her study of smallpox. On her return she engaged in a lively but not too successful campaign to make English doctors see the value of the Turkish method of avoiding severe attacks of what was then the most terrible disease of the age.

MAX MÜLLER (1823-1900) was one of the most celebrated philologists of the nineteenth century, and his lectures and essays went far towards popularizing what had been considered a most difficult subject. In his *Science of Thought*, and other writings of the same type, he opposed the Darwinian theory of the descent of man because he had come to the conclusion that man's command of language, and his power to use it rationally to express his thoughts and to transmit the experience of one generation to those succeeding it, constituted an " impassable barrier " between brutes and men. He studied at Leipzig, Berlin, and Paris, but spent the latter half of his life in England, mainly at Oxford, where he edited a scholarly and comprehensive edition of the sacred books of the East.

SIR GEORGE NEWMAN (1870-1948), Chief Medical Officer of the Ministry of Health from 1919 until 1935, was the leading spirit in the Education Act of 1910 which provided for the medical inspection of school children. He studied medicine at Edinburgh and at King's College, London, where he was appointed Senior Demonstrator of Bacteriology. He was for a time Emeritus Lecturer on Public Health at Bart's, and he served on a large number of committees devoted to the improvement of the nation's health. He received a knighthood in 1911 and a K.C.B. in 1918. His books include *Bacteriology and the Public Health* ; *Infant Mortality* ; and *Interpretations of Nature.*

KENNETH P. OAKLEY (1911-   ), the British Museum expert on fossil man, and author of the handbook on *Man the Tool-Maker*, graduated in geology and anthropology at University College, London, and soon afterwards joined the staff of the Natural History Department of the Museum, where he is now a Principal Scientific Officer. He brings a wide knowledge of other sciences into his geological studies, and has been largely responsible for the recent developments in the method of estimating the age of fossil bones by their fluorine content. His main interest is the borderland where geology and anthropology meet. In 1947 Dr. Oakley visited East Africa, collecting fossils and relics of early man for the Museum, while in 1950 he was guest-lecturer at the Viking Fund Seminar in Physical Anthropology at New York.

LOUIS PASTEUR (1822-95), the founder of the germ theory of disease, and the first to show how many germ-produced diseases could be prevented, had the endless patience of true genius, and a lifelong devotion to hard work. He began his scientific career as a chemist, and his solution of the baffling problem of the optical properties of the tartaric acids earned for him the Chair of Chemistry at Strasbourg University. His researches on fermentation arose out of his first great discovery ; and these led naturally on to his work on bacteria. In 1865 he became famous by his discovery of the cause, and of a method of prevention, of the disease which was devastating the silkworms in France, and in 1880 he achieved the same success with the anthrax which was killing off the sheep and cattle. His last great work was the victory over hydrophobia, and in 1888 the Pasteur Institute was founded, soon to become a place of pilgrimage for thousands of people who had been bitten by mad dogs. Pasteur takes a very high place indeed in the list of benefactors of mankind.

SIR HUMPHRY DAVY ROLLESTON (1862-1944) was typical of all that is best in the British medical service. He studied medicine at St. John's College, Cambridge, and at St. Bartholomew's Hospital. Specializing on the diseases of the liver, gall-bladder, and bile-ducts, he nevertheless took all medicine for his province, and with Sir Clifford Allbutt edited an eleven-volume *System of Medicine*. He was Physician Extraordinary to the King, and sometime President of the Royal

College of Physicians, but also Consulting Physician to the Victoria Hospital for Children. His honours included a baronetcy, a K.C.B., G.C.V.O., and a very large number of honorary degrees. Of interest to the general reader are his *Medical Aspects of Old Age* ; *The Cambridge Medical School* ; and *Aspects of Age, Life and Disease*, from which the lecture quoted in this volume is taken.

SIR E. A. SHARPEY-SCHAFER (1850-1935) was Professor of Physiology, first at London and then at Edinburgh for a total period of sixty-one years. He played a great part in the development of endocrinology, making extracts from the various ductless glands and injecting them into the blood-stream to see what effects were produced. He held degrees from five universities, and became President of the Royal Society of Edinburgh in 1929. In 1931 he invented the Schafer method of artificial respiration.

S. G. SOAL (1889-   ) was the recipient of the first D.Sc. degree ever given by a British university for psychical research. He went to Southend High School, and then to Queen Mary College, London, where he took a First in mathematics and became a Lecturer in that subject. In the First World War he was wounded at Messines, and his youngest brother was killed in 1918. This loss turned his thoughts to psychical research. He resumed his mathematical lectures in 1919, but gave his spare time to the investigation of paranormal phenomena. He was President of the Society for Psychical Research 1949-51, and in the latter year visited America to work in the Parapsychology Laboratory at Duke University. He is the author, with F. Bateman, of *Modern Experiments in Telepathy*, and with H. T. Bowden of *The Mind Readers*.

CHRISTIAN KONRAD SPRENGEL (1750-1816) was a pioneer in the study of botany, a man who made outstanding discoveries because he used his eyes patiently and believed that everything in nature has a reason and a meaning. He was born in Pomerania and studied for the Church, but was ejected from his parsonage in Spandau because he " neglected his flock in favour of flowers ". He went to Berlin where he taught languages and lectured on botany, publishing his great book, *The Secret of Nature*, in 1793.

SIR D'ARCY WENTWORTH THOMPSON, C.B., F.R.S. (1860-1948), was Professor of Natural History at St. Andrews for more than sixty years, but he was a great classical scholar and a mathematician as well as a biologist. He studied medicine at Edinburgh for two years, but then won a scholarship to Trinity College, Cambridge, where he read zoology and was appointed Demonstrator in Physiology. In 1884 he took the Chair of Natural History at University College, Dundee, and retained this post when the college was incorporated into St. Andrews University. Much of his time was given to fishery research and regulation, but he found time for classical studies as well, and he was President of the Classical Association in 1929. His main technical work is *Growth and Form*, in which the shapes of living things are discussed and explained. To commemorate his sixty years as a professor a number of prominent biologists collaborated to produce a presentation volume : *Essays on Growth and Form*. A reading of his lecture will show that he had just as keen a feeling for form in sentences as in animals.

SIR LANDSBOROUGH THOMSON, C.B. (1890-   ), son of the well-known writer on popular science, Sir J. Arthur Thomson, has devoted most of his life to the Medical Research Council, to which he became Assistant Secretary in 1919, and then, in 1949, Second Secretary. He has been Chairman of the Public Health Laboratory Service Board since 1950. He went to school at Edinburgh and Aberdeen, and studied in the universities of Heidelberg, Aberdeen, and Vienna. He served with the Argyll and Sutherland Highlanders in France and Flanders in 1915-19, and attained the rank of Lt.-Colonel, being mentioned in despatches and awarded the O.B.E. Birds are his hobby. He was President of the Ornithologists' Union, 1948-55. He was also President of the eleventh International Ornithological Congress. His publications include *Problems of Bird Migration*; *Birds, an Introduction to Ornithology*; and a very popular *Bird Migration : a Short Account*.

GEORGE EDGAR VINCENT (1864-1941), President of the Rockefeller Foundation of Hygiene in New York from 1917 to 1929, was born in Illinois and took his Ph.D. at Chicago, where he was Fellow in Sociology in 1892, and Professor of Sociology from 1904 to 1911. He then went to the University of Minnesota as President. He received honorary degrees from Chicago, Yale, and Michigan.

C. H. WADDINGTON, F.R.S. (1905-   ), is Buchanan Professor of Animal Genetics in the University of Edinburgh, and Hon. Director, Agricultural Research Council's Unit of Animal Genetics. After taking a First in Geology he turned his attention to biology, holding travelling Rockefeller Foundation fellowships in 1932 and 1938, and being Lecturer in Zoology and Embryologist in the Strangeways Research Laboratory at Cambridge from 1933 to 1945. He carried out operational research for Coastal Command of the R.A.F. in 1942-45, and was Scientific Adviser to the C.-in-C. during the last two years of the war. His publications include *Introduction to Modern Genetics*; *Organizers and Genes*; and *The Strategy of the Genes*.

SIR J. A. SCOTT WATSON (1889-   ) was Chief Scientific and Agricultural Adviser to the Ministry of Agriculture, and Director-General of the National Agricultural Advisory Service, 1948-54. The son of a farmer near Dundee, he went to the universities of Edinburgh and Berlin, and then to the Iowa State College of Agriculture. He lectured for a time in the Department of Agriculture in Edinburgh University and then went to France with the R.F.A. in 1915, winning the M.C. He occupied the Chair of Agriculture at Edinburgh, 1922-25, and that of Rural Economy at Oxford, 1925-44. He was Agricultural Attaché to the British Embassy at Washington, 1942-44, and Editor of the Journal of the Royal Agricultural Society, 1931-45. He was knighted in 1949. His publications include: *Agriculture—The Science and Practice of British Farming*, a standard text-book written with J. A. More; also *Rural Britain To-day and To-morrow*; *Great Farmers*; *The Farming Year*; and a *History of the Royal Agricultural Society*.

NORBERT WIENER (1894-   ), Professor of Mathematics at Massachusetts Institute of Technology since 1932, has sponsored the new science of "cybernetics", the study of control and communication in animals and in machines. He graduated from Tufts College at the age of 15, and received a Ph.D. from Harvard at 19. He continued his studies at Cornell and Columbia in America, and then crossed the Atlantic to attend Cambridge, Göttingen, and Copenhagen. His achievements in mathematics have brought him international fame, and they were put to good use in solving problems of fire control and radar during the second world war. His book *Cybernetics*, first published in 1947, caused a sensation in scientific circles, and as it possessed a "rather forbidding mathematical core" he was persuaded to write a related book for lay reading. Hence *The Human Use of Human Beings*, from which our extract was taken.

SIR SOLLY ZUCKERMAN, C.B., F.R.S. (1904-   ), Sands Cox Professor of Anatomy in Birmingham University, is Deputy Chairman of the Advisory Council on Scientific Policy, a member of the Agricultural Research Council, and of many other learned bodies. He was born in Cape Town, and after studying medicine in the university there he came to London as Research Anatomist to the Zoological Society, and

Demonstrator in Anatomy in University College. In 1933-34 he worked in America as Rockefeller Research Fellow, and then took an appointment as Demonstrator and Lecturer in Human Anatomy at Oxford, 1934-45. He held the rank of Group Captain (Hon.) with the R.A.F. from 1943 to 1945, and was awarded the American Medal of Freedom with Silver Palm. He is editor of the *Journal of Endocrinology ;* and in addition to a great nember of technical and scientific papers he has written *The Social Life of Monkeys and Apes* ; and *Functional Affinities of Man, Monkeys and Apes.*

BIBLIOGRAPHY

HERE are the titles of a few books which can be recommended for further reading :

Wells and Huxley. *The Science of Life.*
John A. Moore. *Principles of Zoology.*
F. H. Shorsmith. *Life in the Plant World.*
R. M. Buchsbaum. *Animals without Backbones.*
J. Z. Young. *The Life of Vertebrates.*
W. H. Thorpe. *Learning and Instinct in Animals.*
Sir Julian Huxley. *The Uniqueness of Man.*
Alan Dale. *Social Biology.*
C. H. Waddington. *The Strategy of the Genes.*
Norbert Wiener. *The Human Use of Human Beings.*
" *Scientific American* Books." *The Physics and Chemistry of Life* and *Twentieth Century Bestiary.*
Penguin Books Ltd. *New Biology* (Periodical).

ACKNOWLEDGMENTS

FOR permission to include copyright material in this volume our thanks and acknowledgments are due to the following authors and publishers :

To Dr. Michael Abercrombie for " From Egg to Adult ", which first appeared in *New Biology*, No. 5 ; to Dr. C. H. Andrewes for " Can Influenza be Reduced ? " ; to the editor of the *British Medical Journal* for Lord Dawson's " Medical Science and Social Progress " ; to the late Professor Dendy for " The Stream of Life " and " Biological Foundations of Society " ; to Prof. D. W. Ewer for " The Biology of Enzymes " ; to Dr. Eggleton for " The Chemistry of Life " ; to Professor John E. Harris for " Structure and Function in the Living Cell " ; to Professor A. V. Hill for " Experiments on Frogs and Men " ; to the Very Rev. W. R. Matthews for " Man as Science sees Him " ; to the late Sir George Newman for " Public Opinion in Preventive Medicine " ; to Dr. K. P. Oakley for " A Definition of Man " ; to the late Sir Edward Sharpey-Schafer for " The Human Post Office " ; to Dr. S. G. Soal for " Seeing into Future Time " ; to the late Sir D'Arcy W. Thompson for " The Greater Problems of Biology " ; to Sir Landsborough Thomson for " Why do Birds Migrate ? " ; to the late Dr. G. E. Vincent for " International Public Health " ; to Professor C. H. Waddington for " The Biological Order " ; to Sir J. A. Scott Watson for " The Pressure of Population on Land Resources " ; and to Professor Sir S. Zuckerman for " What is a Hormone ? "

We are indebted to Messrs Routledge and Kegan Paul, Ltd., and to the authors for permission to include " How Life Began ", from Professor J. D. Bernal's *The Physical Basis of Life*, and Sir Humphry Rolleston's " Concerning Old Age ", from *Aspects of Age, Life and Disease* ; to Sir Julian Huxley and Messrs. Chatto & Windus for " Genetics as a Science ", from *Soviet Genetics and World Science* ; to Messrs. Eyre & Spottiswoode, Ltd., and the author for Professor Norbert Wiener's " Ants and Men ", from *The Human Use of Human Beings* ; to Dr. E. B. Ford and the Clarendon Press for " The Genetic Basis of Adaptation ", from *Evolution : Essays Presented to Professor Goodrich* ; to the Cambridge University Press for " Experiments in Plant Hybridization ", from *Mendel's Principles of Heredity*, by W. Bateson. We are indebted to Sir Alexander Fleming, to the Council of

St. John's College, and to the Cambridge University Press for permission to include the extract from Sir Alexander's Linacre lecture on Chemotherapy ; and to the authors and to Messrs. George Allen & Unwin, Ltd., for Sir Julian Huxley's "Progress in the Evolutionary Future", from *Evolution: The Modern Synthesis*; and G. Dahlberg's "Genetics and Eugenics", from *Race, Reason and Rubbish*.

Indulgence is craved for any inadvertent failure to acknowledge help received.

INDEX

(Titles of Lectures are in Black Type)

348 BIOLOGY